Harnessing The Sun

A practical no-nonsense approach to home solar heating

by
John Keyes

MORGAN & MORGAN, Publishers
Dobbs Ferry, New York

First Printing, December 1974
Second Printing, April 1975
Third Printing, August 1975

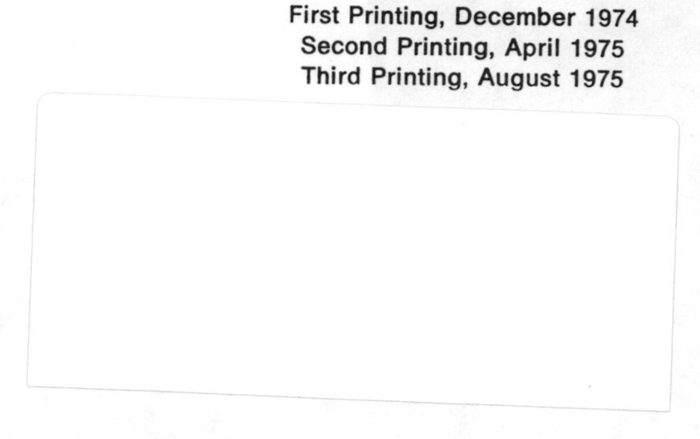

International Standard Book Number 87100-089-X
Library of Congress Catalog Card Number 75-7852

Printed in U.S.A.

ACKNOWLEDGMENTS

The author gratefully acknowledges the patient and careful analysis by Dr. Harold Svanoe, Harry Remmers and the entire staff of International Solarthermics Corporation (ISC) during the preparation of this book; the support and enthusiasm of Lester and Doris Nelson, Erika, Betty and many others during its infancy; and most of all, the hard-working ISC backyard solar furnace which made it all possible.

INTRODUCTION TO 2ND EDITION

Very little has changed in the field of Solar Heating Applications since the writing of this book in 1974. Unfortunately, it is still the case that much of the governmentally funded research is "paper research" with the emphasis on components of systems as opposed to practical, down-to-earth engineering of complete systems. *There is no mystery about solar heating systems!* Hopefully this introductory book to Solar Heating Applications will serve to dispel the totally erroneous idea that solar furnaces must be esoteric "Buck Rogers" devices in order to function.

One thing has changed in the field of Solar Energy, however: the blue-suede shoe boys are flocking in droves to this field, smelling a "quick buck" to be made. Hundreds of "solar energy consultants" have sprung into existence ready for a fee to give you advice about the system you are going to build. Unless they have a registered professional engineer licensed in your state on their staff, run, do not walk, out the door! A trip to the tables in Las Vegas would give you more of a chance!

Reminiscent of the uranium days, 1¢ and 10¢ stocks are abounding in newly formed solar companies. Although this alone is not damning, investigate before you invest! The author recently received a *form letter* from a stock broker offering to prepare such an offering of 10¢ stock with a promise of "just give us a call so we can have your $500,000 to you in two months' time!"

Solar heat is practical and economical NOW. You need have no fear of installing a solar system manufactured by a reputable firm, one which clearly states warranty and performance information about its products.

Appendix G contains some common-sense tips to anyone considering the purchase of a solar heating system. When you consider solar heat, please use that checklist to protect yourself from the get-rich-quick boys!

The fuel situation is worsening much more quickly than expected, and the oil and gas import balance of payments degrades daily.

With inflation the way it is, do not expect prices for systems to go down: they will go up, particularly in the face of increasing demand in future months and years.

Manufacturers all over the United States are producing or gearing up to produce factory units. The Solar Age has dawned!

John Keyes
March, 1975

FOREWORD

This book was prepared to give the reader a general knowledge of the field of Solar Energy Applications and was written as objectively as possible. It has been the intent of the author to present *practical* information in easily readable form, as opposed to lengthy scientific dissertations.

The author is one of the originators and inventors of the first practical self-contained solar furnace auxiliary home heating system now being manufactured and marketed under a variety of trade names by licensees of International Solarthermics Corporation of Nederland, Colorado. Many references to that unit will be found herein, as well as comparisons to other systems. Hopefully, such references will be illustrative or educational in nature as opposed to being blatantly commercial. So far as is known, this solar furnace is the only practical, compact, backyard solar furnace being factory-produced for widespread residential application anywhere in the world as of this date.

Even so, it is but the forerunner of many practical mass-produced units which will most certainly be produced in the future. Perhaps this practical handbook will inspire young solar researchers to build upon the basic principles contained herein to design ever-better and more-efficient systems.

More importantly, by acquainting the reader with the processes involved in solar heating, it is to be hoped that he may sift through the various theories with which he will be bombarded in the coming years and common-sensically evaluate which of them are practical. And practicality is the key word in this exciting new field. No matter how esoteric and technologically impressive a new design may be, the real crux of the situation is: 1. Does it work? and 2. How much will it cost compared to what it replaces?

Support the young men in your community who are expending their time, money, and efforts to find better ways of harnessing the sun. They've rolled up their sleeves and are seeing what works in practice. It is from these efforts that the science of Solar Energy Applications will progress and not from lavishly funded government research projects which decide in advance what will work and what will not.

Further, the reader is urged to keep a scientific attitude while evaluating any system he may encounter in the days to come. As might be expected, many self-proclaimed "experts" are already busy explaining from their armchairs why this or that system cannot work from a theoretical standpoint, instead of examining the system in operation to see if it does or does not work!

Without question, however, the age-old rule of *Caveat emptor* – Let the buyer beware! – prevails. Investigate thoroughly before you invest in any solar heating system for your home!

John H. Keyes
May 1974

CONTENTS

CHAPTER 1

SOLAR HEATING APPLICATIONS

The energy crisis has given birth to a tremendous interest in the use of solar energy as a possible solution. As a result, literally hundreds of firms and individuals have set themselves up as experts — able to dispense advice on solar engineering — after a cursory study of the available literature in this field.

The science of Solar Heating Applications is in its infancy! Appallingly little *practical* research has been done, in light of the large number of "basic principles of solar heating" being promulgated by solar experts. A great quantity of the research being done in this field is "paper research" — research done from a theoretical standpoint — utilizing the physical laws from other scientific disciplines which many times have little or no practical application in the actual utilization of solar energy.

An example of "paper research" is the examination of the aerodynamic principles involved in the flight of the bumblebee. The conclusion of such a paper study is that a bumblebee can't fly — it is aerodynamically impossible. Obviously, a practical man forms a different conclusion. But this demonstrates the dangers of forming conclusions based upon "paper research" without confirming those conclusions in the real world of experience.

It can be categorically stated that, at this point in the progress of solar energy applications, *there are no experts* (including the author) in this field. There are only researchers.

THE PRACTICAL APPROACH TO APPLICATIONS

As might be expected in a new field, the cart is put before the horse more times than not. That is, a system is designed with the hope that, when it is installed, it will work! This approach, of necessity, makes every installation a prototype which may or may not work.

From a practical standpoint, a given system should be built, tested and evaluated FIRST on a typical type installation. THEN, it should only be installed on similar type structures to the one on which the unit is known to work. For example, several prototype models of the International Solarthermics Corporation (ISC) backyard solar furnace were built, tested and evaluated in use. As a result of this work, it was concluded that this unit has a *specified* application. It is designed for residential use on very well insulated three or four bedroom homes built in the past fifteen years in suburban and rural areas. Since that description can cover widely varying types of homes, a more exact specification is to limit its use to homes ranging in size from 5,000 to 15,000 Btu/Degree Days in heating requirements. Obviously, its efficiency varies considerably within that range. (For an explanation of Degree Day house sizes, see Chapter Six). For a home exceeding 15,000 Btu/Degree Days, two units might be required to provide the efficiency desired. At present, three models of this unit are being marketed. Any installation which does not conform to the design parameters described above would, of necessity, be a prototype installation.

There are, of course, other firms and individuals utilizing this practical approach. Before buying a given unit, or paying for advice about a given installation, it is advisable to ask the following practical questions:

1. How many *working* systems of this sort have you designed for specifically this size and type of installation?

2. What was the total cost of that installation, and how long ago was it installed? (Construction prices have been sky-rocketing, and a 1958 price, for example, would be downright deceptive. Be sure that all labor, all excavation, all materials and a reasonable construction profit have been included.)
3. Are the costs separated from house construction costs? (Many times a roof-top collector, which requires extraordinary roof angles with substantial increases over a typical roof cost, will be priced sepa-

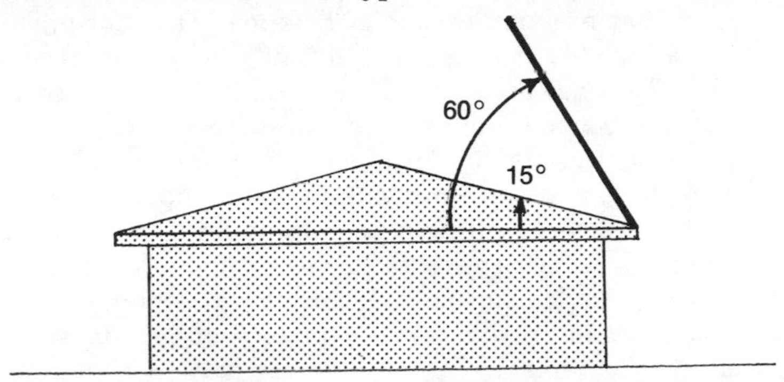

Figure 1. Roof-top collector installed at 60° angle represents tremendous increase in cost of construction compared to typical 4:12 pitch roof. This additional cost is attributable directly to the cost of the solar heating system.

rately — "You have to have a roof anyhow." — listing only the collector materials as part of the cost. Similarly, basement heat storage cost computations will ignore the fact that concrete to contain

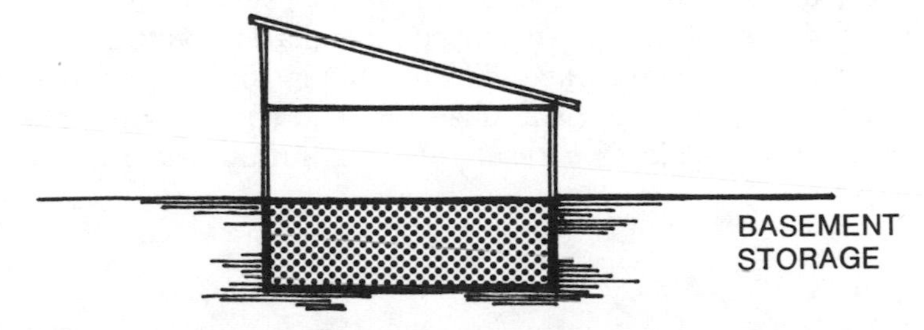

Figure 2. The portion of the basement occupied by storage must be included in the cost of storage; i.e., costs of excavating, forming, concrete, etc. must be pro-rated: part to the house, part to the solar heating system.

storage costs money, and that a substantial reduction in living space is effected – "You have to have a basement anyhow." Insist on a total comparative cost – one which shows an ordinary house yielding exactly the same amount of living space with ordinary construction practices and the solar house which must be larger and have less usable space. This kind of accounting for the cost of installation will be vastly different!)

4. What will the overall efficiency be? (Honest installers will tell you right away that from an economic standpoint it is not practical to build a unit capable of handling 100% of your heating needs. The collector and storage area are expensive to build and install, and it doesn't make economic sense to add to these expenses for a capability which is utilized only during extreme weather and sun conditions. Therefore, most systems compromise economics and practice to provide an average heating capability under usual conditions, or some reasonable percentage thereof, and rely upon a fueled or electric system to provide backup heat during the extreme periods. Most systems will try to provide from 40 to 85% of the annual heating requirements from solar energy.)
5. What kind of maintenance and upkeep will it require?

If you are considering the installation of a solar heating system which can only be categorized as a prototype installation, protect yourself. Read very carefully the warranty you are going to be provided, and determine how secure the company or person backing it is. It is the opinion of the author that such prototype research should be encouraged, if you have the money to risk on such an investment. However, if you are funding a prototype installation, you have every right to insist upon a percentage of ownership in that design, particularly if patents are going to be applied for. You should participate in any future profits made as a result of your initial risk investment!

CHAPTER 2

THE SUN

The sun is, of course, the source of fuel for all solar energy applications. It puts forth incomprehensibly large quantities of energy every minute. Even though only miniscule portions of this energy ever reach Earth, the amount that does arrive, properly harnessed, is more than enough to supply *all* of man's needs! More importantly, this energy is inexhaustible, and the use of it is non-polluting.

The energy which arrives from the sun is in the form of radiation. In solar heating applications, we are particularly interested in heat energy, and we measure it typically in "Btu's" (British Thermal Units). A Btu is the amount of heat energy required to raise the temperature of one pound of water one degree Fahrenheit. For example, to relate the Btu to ordinary life, consider that in Denver, Colorado a 10,000 Btu/Degree Day house typically insulated will require about 365,000 Btu's of heat energy to keep the inside temperature at 70°F for an average 24 hour period in January.

Other measuring units are sometimes used for solar energy, such as calories, langleys, joules or ergs. In practice, most ordinary furnaces are rated in Btu's, and most heating contractors do their heat loss calculations on a given structure in Btu's, so this is the unit of measurement that you will ordinarily encounter. Don't be misled into thinking that because a given application is rated in one of the other units of measurement that it is in some way more esoteric or superior.

THE SOLAR CONSTANT

A figure utilized in calculating the amount of solar radiation (or *solar insolation)* reaching Earth is the *Solar Constant.* (As a matter of fact, the radiation itself is not constant at all, but varies considerably from time to time.) The Solar Constant is taken at a point outside the Earth's atmosphere at a distance of the diameter of Earth from the surface. Approximately 442 Btu's fall on a square foot of surface area perpendicular to the sun's rays in the period of one hour. This is usually written as 442 Btu/ft²•hr. In solar heating applications, however, the figure for the Solar Constant has little applicability, since very few dwellings are located outside the Earth's atmosphere!

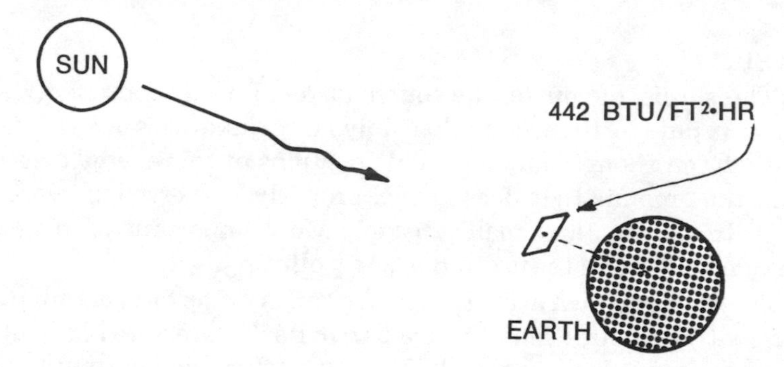

Figure 3. A one square foot plane placed normal to the sun's rays outside of the earth's atmosphere will receive approximately 442 Btu's per hour.

From a practical standpoint, you are concerned with just how much of this energy strikes the Earth's surface in the geographical location where you live. Unfortunately, this data is available only for a few cities. Indeed, only in the past few years has any concerted collection of this data been performed.

As radiation from the sun travels through the atmosphere, quantities of it are absorbed by air molecules, dust particles and water droplets in the air, to be radiated again. This re-radiation is omnidirectional, so that only a part of it is re-radiated to the Earth's surface, while most of it is re-radiated to outer space. The portion of this re-radiation and reflected radiation which eventually reaches the Earth's surface is called *Diffuse Radiation, Indirect Solar Radiation*, or *Sky Radiation*. The portion of the heat energy which penetrates the atmosphere without being absorbed is called *Direct Solar Radiation* or *Direct Solar Insolation*.

This heat energy which arrives on Earth from the sun is in the form of radiation. Only a small portion of this radiation is visible to human eyes, and that part is called "light." Some portion of all visible light which falls upon objects is absorbed and re-radiated as heat. Thus the visible portion of the radiation spectrum is very important in solar heating applications. Radiation from the sun travels through space to Earth at the fantastic speed of 186,000 miles per second. Even so, it takes over eight minutes for this light to reach us from the sun, over 93,000,000 miles away!

PRACTICAL UTILIZATION OF SOLAR ENERGY

As you know, the amount of visible light arriving at Earth during daylight hours is controlled by weather conditions. On a dark, overcast day very little of this light penetrates the clouds.

This is one of the major barriers to harnessing the sun's energy, because in utilizing that energy practically, we must have that energy on tap – available for use at all times. But the fuel input to a solar collector varies considerably, and it is not always intense during the time that we require large quantities of energy. Additionally, it is not present at all during the night time – just when energy demands are the highest. Therefore, any practical solar heating device must not only efficiently collect the sun's heat, but more importantly, it must *store* heat for use at times when sunshine is not present. So it is very important, when evaluating a given solar heating application, to pay close attention to the provisions made for storage and their practicality.

THE WAVELENGTH CONVERSION PROCESS

A process which takes place as heat energy arrives from the sun, and with which you must be familiar to understand the operation of a solar heating system, is *wavelength conversion*. Contained within the thermal (heat) spectrum of radiation are two relatively distinct spectrums. The direct solar radiation arriving from the sun is in the form of shortwave radiation (0.3 to 3 microns in length). After this shortwave radiation strikes matter and is absorbed, however, it is re-radiated in longwave form (3 to 30 microns in length).

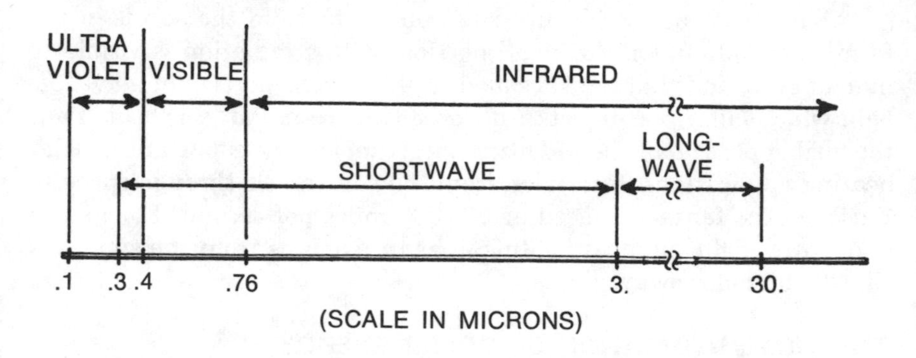

Figure 4. Diagram of the portion of the electromagnetic spectrum used in solar heating devices, showing the shortwave and longwave portions of the thermal spectrum.

Very little is known about this wavelength conversion process: why it happens, the dynamics of the change, etc. Yet this wavelength conversion process is all-important to the efficient operation of a typical solar collector. Which points out just how little is really known and understood in the field of Solar Energy Applications!

CHAPTER 3

THE SOLAR COLLECTOR

Literally hundreds of different configurations have been tried in the quest to capture energy from the sun. Every type of collector has certain advantages, and usually, a large group of accompanying disadvantages. One could say, with a certain amount of truth, that research in the area of solar collectors has been aimed at minimizing the disadvantages!

Collectors may be arbitrarily divided into several categories: 1. Electrical Conversion, 2. One-Step Thermal, 3. Concentrating Thermal, and 4. Flat-Plate or Stationary-Plate Thermal.

ELECTRICAL CONVERSION COLLECTORS

These collectors utilize special solar cells which convert solar energy into electricity. The significance of such devices is considerable, making research with them very exciting. Solar cells have been utilized extensively in the space program and have been used for some time by the telephone companies to power repeater stations along rural transmission lines.

The University of Delaware has built a solar heated *and* powered home ("Solar One") utilizing solar cells which promise to be less expensive and more dependable than the first solar cells.

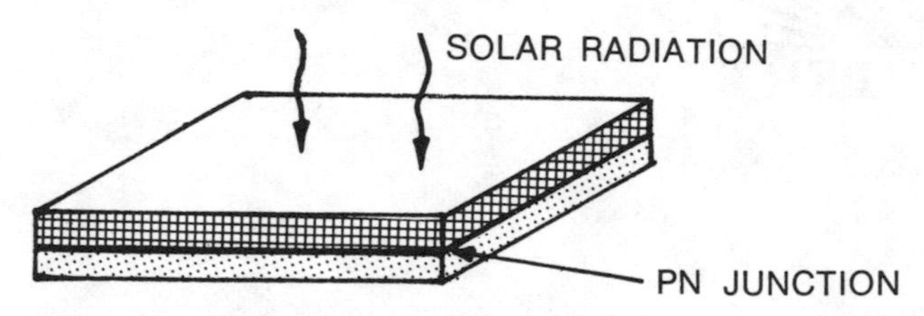

Figure 5. Schematic drawing of typical solar cell. Dissimilar materials are bonded together. Excitation by solar radiation causes electrons to cross the PN junction, and electrical flow is created.

Present prices of solar cells are exorbitant, with the cost per kilowatt at about $3000. For an all-electric home using 100 kilowatts per day, the cost of installation would be in excess of $300,000 per residence. The new, less expensive cells could bring this down to as low as $30,000 per residence, which might be within the realm of possibility if it were not for a second drawback: the life expectancy of the new cell is only about five years, making the *annual* maintenance of the system about $6000!

Most probably there will never be widespread residential use of solar cells.

ONE-STEP THERMAL COLLECTORS

One-step thermal collectors are just what the name implies. In

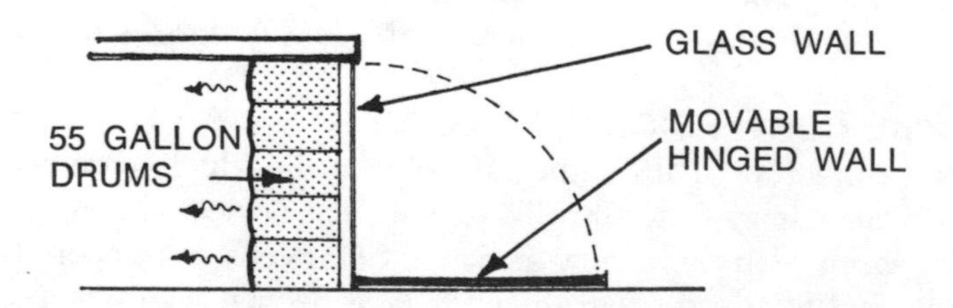

Figure 6. One-step thermal collector. Relatively inexpensive, it is a good example of the creative innovation occurring in the field of solar energy applications today.

general, these are uncontrolled solar heating devices. An example of such a collector is a device utilizing a bank of 55 gallon drums filled with water and painted black. This bank of drums is located behind a glass wall on the south side of the house. During sunshine time, these drums are heated. At night, a hinged wall is raised to cover the glass wall to help prevent the loss of heat collected during the daytime. During heating phases, heat is also being transmitted to the interior of the structure, which may or may not be desired. Heating at night is also uncontrolled in such a system, with the heat being provided possibly being too much or too little. However, such a system is relatively inexpensive!

The major obstacle to the introduction of such one-step thermal devices in a widespread way is the fact that the system may require a major change in the life habits of the occupants. Although there are some people willing to adapt to such changes, the majority of people prefer not to change, which means that the system has to be changed to accommodate the living habits of the average American.

CONCENTRATING THERMAL COLLECTORS

Concentrating thermal collectors are devices which reflect or refract incoming direct solar radiation to a specific focal point. They are useless with diffuse radiation for this reason. Generally, they are in

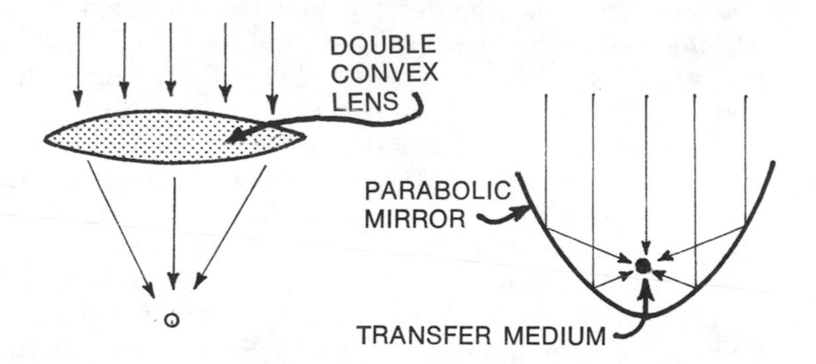

Figure 7. Two types of concentrating collectors. The double convex lens is the familiar magnifying glass. The parabolic mirror is used more often in solar heating applications.

the shape of a parabolic mirror or a double convex lens. Tremendously high temperatures can be achieved with a concentrating collector. Indeed, many are on the market now for use as ovens or barbecues because of this high temperature capability. You have used a concentrating collector if you have started a fire with a magnifying glass.

The major practical problem associated with concentrating collectors is that they must be aimed directly at the sun in order to function. Unfortunately, the track that the sun follows from sunrise to sunset is different every day of the year. Therefore, extremely complex and expensive equipment must be used to keep the collector aimed directly at the sun. From an economic standpoint it is simply not presently practical to produce concentrating collectors for residential use. A sophisticated tracking device could cost more than the house itself!

Structural problems require that the cost of such a system be quite high. Further, the high temperatures produced are simply not necessary in residential applications and require unnecessarily sophisticated methods be utilized in storage of heat energy – again raising the price of the system out of reach of the ordinary man.

FLAT-PLATE OR STATIONARY-PLATE THERMAL COLLECTORS

Because of the problems associated with the collectors mentioned above, most research has been devoted to perfecting the flat-plate or stationary-plate thermal collector. These fall into three broad categories: 1. Single Fixed Plate, 2. Complex Fixed Plate, and 3. Vertical Fixed Plate.

Before examining each of these categories individually, it is important to examine the location of typical collectors. Nearly all systems use a *roof-top mounting*. Why? Because with most systems, *huge* collectors are required. Many have collector surfaces ranging from 1000 square feet (20 feet by 50 feet) to 2000 square feet (36 feet by 56 feet) in area. With a surface this big, the roof is a good place to put it to get it out of the way. By comparison, the ISC backyard solar furnace requires a collector surface of only 96 square feet (8 feet by 12 feet) and thus can be installed without appreciable loss of usable ground space in a backyard.

This consideration becomes extremely important when one is contemplating adding solar heating to an existing residence, particularly when one comes to the next requirement — that the collector be installed at a 60° angle. A typical roof on an existing residence is inclined at only about a 15° angle. It is practically impossible to add a 60°-pitched collector to a 15°-pitched roof without starting over from scratch! Wind loads and weight stresses are just too great for roof trusses which were designed for totally different conditions.

Thus, systems utilizing large roof-top collectors are used in most cases only in new construction, with the house being built around the solar heating system.

A second practical problem to utilizing a roof-top collector is that the heat energy captured must be transferred great distances to the storage area, and transmission losses can be appreciable.

STRUCTURE OF FIXED-PLATE COLLECTORS

The structure of fixed-plate collectors is very similar. Each is comprised of five parts as shown in Figure 8 below.

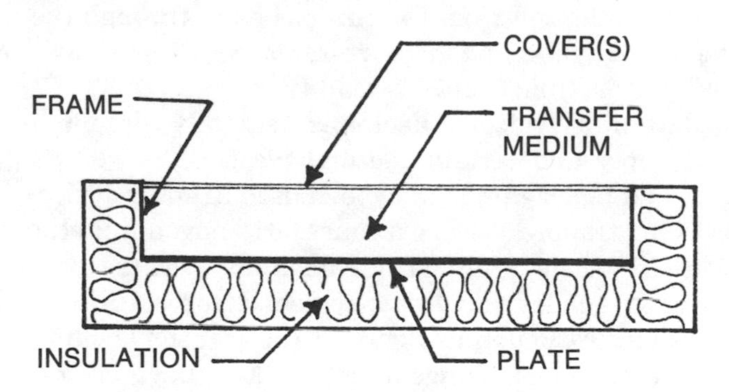

Figure 8. Components of fixed-plate collectors.

As can be seen, the entire collector is in the shape of a big, shallow box with insulation surrounding its perimeter and covering its back. The cover is comprised of one or more layers of glass or plastic and is extremely important to the functioning of the collector. As mentioned earlier, heat energy from the sun arrives in the form of shortwave radiation. It is a property of glass, and of some plastics, that they are relatively transparent to this shortwave radiation, allowing it to pass on through and strike the plate. The plate is coated with a flat black paint which absorbs most of this shortwave radiation. After being absorbed by the plate, this energy is re-radiated as longwave radiation. It is another property of glass to be virtually opaque to longwave radiation.

In rough figures, about 90% of shortwave radiation will pass through a single pane of glass, while no longwave radiation will pass through that same glass. Thus, with two glass covers separated by a small air space, about 80% of the incoming shortwave radiation will pass through to strike the black plate where approximately 95% of it will be absorbed and 5% reflected back through the covers.

In this manner, the fixed-plate collector *traps* the heat energy from the sun. You have probably encountered this effect in ordinary life. Even though it is cold outside, you can get into a car on a sunny day in winter time and the inside temperature will actually be hot! The shortwave radiation from the sun has gone through the windows, struck the seat covers, and been re-radiated as longwave radiation which didn't escape from the car as readily.

The plate of a typical collector is generally aluminum, painted black; but copper and certain thermal plastics, as well as graphite-coated glass and black gauze, have been used in some devices.

After being trapped, this heat must be removed from the collector. This is done by means of a transfer medium. If water or some other liquid is used, it is called a *hydronic* system. If air is utilized, it is called a *hot air* system. Although in Figure 8 the transfer medium is shown above the plate, dozens of ingenious methods have been devised for removing captured heat from the collector: above, below, through and around the plate itself.

THE SINGLE FIXED-PLATE COLLECTOR

The typical single fixed-plate collector has one or more covers over an insulated box at the back of which is a plate. Practice has shown aluminum, due to its very high thermal conductivity, to be a good and

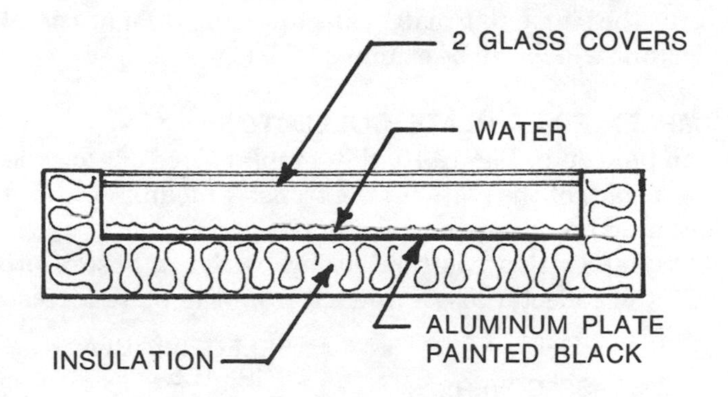

Figure 9. A typical single fixed-plate collector.

relatively inexpensive plate material. Gold and silver would also be good plate materials, but cost obviously rules them out.

Glass turns out to be the most practical cover. Acrylic plastic sheets, because of their extremely high coefficients of linear expansion, are not as practical. It is almost impossible to glaze and seal plastic covers properly, unless the collector doesn't operate at very high temperatures.

Notice that, in the single fixed-plate collector, the plate serves as the absorption surface as well as the heat transfer surface. This dual utilization of the same surface causes real problems if water is used as the transfer medium.

No matter how well sealed the system is, in the real world dust collects on the plate. This dust begins to build up and interfere with collection on non-washed surfaces and forms a scale on the washed surfaces. This build-up increases the reflected losses of the shortwave energy back out of the collector.

The best example of the single flat-plate collector is the Harry Thomason system, which uses corrugated aluminum sheeting as the plate with water running down the roof during collection. This effective and simple application devised over 18 years ago has one major drawback: it makes good common sense and is not terribly "technological." Therefore, tremendous amounts of money have been spent "improving" the single flat-plate collector using dual or complex fixed-plate collectors, which will be examined next.

THE COMPLEX FIXED-PLATE COLLECTOR

As can be seen in Figure 10, the complex fixed-plate collector uses a separate surface of the plate for the transfer medium – different from the surface used for absorption.

Unfortunately, this "solution" brings with it a host of other problems. First, the collector losses upward from the plate increase signifi-

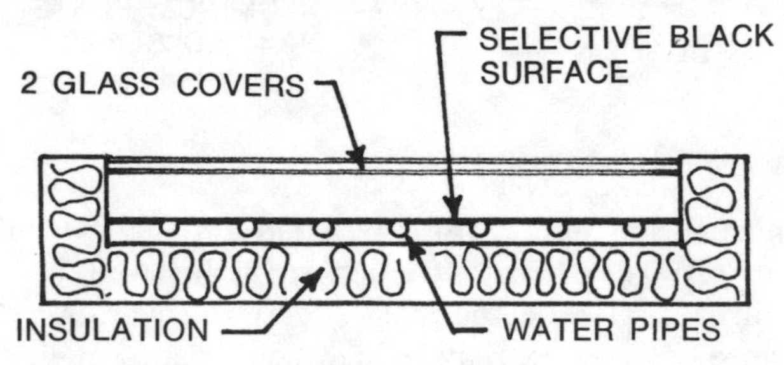

Figure 10. A typical complex fixed-plate collector.

cantly. The solution was a special "selective black" paint which permitted the entry and absorption of shortwave radiation, but interfered with the re-radiation of longwave radiation. The paint is relatively expensive, but does a good job.

The next problem is in effecting heat transfer from the other surface of the plate. As can be seen in Figure 11, many different con-

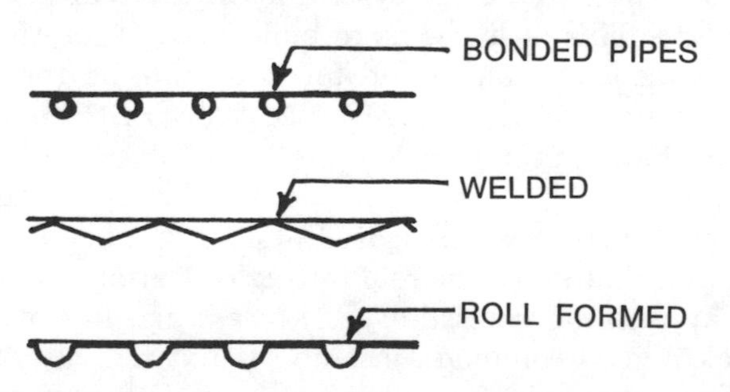

Figure 11. Some varieties of complex fixed-plate collectors.

figurations have been tried. Nearly all of these require, however, a process known as "thermal bonding". Unless a good thermal bond is achieved heat transfer is very poor, and unless the heat collected is effectively transferred the collector is not much good. Unfortunately, good thermal bonding is difficult to achieve and is very expensive in labor and materials.

Despite these practical drawbacks, the complex fixed-plate collector is very satisfying technologically — no longer is it a simple device, but one which requires special fabrication techniques. In short, it looks more complex and more scientific. Little matter that it may or may not work as well as simpler devices. In fact, a recent study by the University of Pennsylvania showed very little, if any, improvement upon the performance of a typical flat-plate collector by many esoteric "improvements" in design!

THE VERTICAL FIXED-PLATE COLLECTOR

The most exciting discovery in recent years in the field of solar energy applications has been the invention of the vertical fixed-plate collector utilized on the ISC backyard solar furnace.

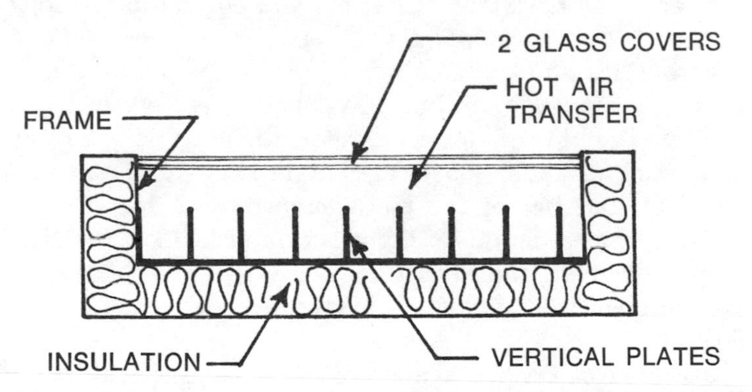

Figure 12. The vertical fixed-plate collector.

Although no more energy impinges on the vertical fixed-plate collector than on a single-plate collector of the same size, significant increases in collection and output capability are observed. Further, heat losses from the collector are decreased with relation to the quantity of solar energy collected.

By the use of some 1872 aluminum cups as vertical vanes, an interior surface area exposed to the transfer medium in excess of 675 square feet is achieved beneath a cover having only 92 square feet of surface area, after subtracting for glazing supports.

The action of the transfer medium in this collector is very important to its efficiency. The transfer medium is a relatively high pressure air stream which is designed to create high turbulence between and above each of the vertical vanes. This turbulent air stream thus "washes" the accumulated heat from the vanes for transfer to the heat storage area.

From a theoretical standpoint, the additional surface area inside the collector should have no effect, since the "window" of the cover is the controlling factor to the amount of energy impinging upon the plates. In practice, however, even very good flat black paints are not perfect absorbers, so that significant portions of the shortwave radiation passing through the cover to a typical flat plate are simply *reflected* right back out of the collector! The vertical vanes, however, tend to reflect this shortwave radiation, as well as trapped longwave radiation, to another surface *inside* the collector. The resulting heat is removed quickly and efficiently by a carefully engineered airflow which keeps the airstream and plate temperatures very nearly equal.

Radiative heat losses upward are significantly less for this collector because of its smaller cover area and cooler cover temperatures. The flow characteristics of the transfer medium across the vertical vanes are such that a boundary layer is established next to the inner cover, thereby reducing the cover temperatures. Similarly, the conductive losses through the smaller cover and sidewall areas are less for this collector.

Because of its capability to maximize heat collection and minimize heat losses, a small vertical fixed-plate collector can capture as much usable heat energy as a much larger, less efficient flat-plate collector.

SURROGATE COLLECTION

It is an economic fact of life that collectors are quite expensive to produce. Even simple collectors will cost from $8.00 to $10.00 per square foot of collector area, if one includes labor, materials and construction profit. Why? Look at it common-sensically. Two panes of glass alone will cost about $2.00 per square foot. Look at a window in your home. The *glass* is the least expensive part of the window! The frame to hold it costs much more. This means the cover glasses alone of a collector will cost $6.00 per square foot or more installed, when labor and materials and profit are accounted for! Then the cost of the plate and insulation must be added. No amount of optimistic cost accounting can get around this simple economic fact. Collectors are expensive on a per-square-foot basis. Therefore, from a design standpoint, it is important to keep the collector area as small as possible and still produce sufficient output for heating purposes.

One way of boosting collection capability is through the use of reflective surfaces which direct additional heat energy into a collector by aimed reflection. Such a reflective surface costs much less per square foot than does the collector, but has the practical effect of *effectively* increasing the size of the collector by increasing the amount of radiation reaching the collector.

The ISC backyard solar furnace utilizes such reflective surfaces, but gets dual use from them. During the winter months they lie parallel to the ground, serving to increase collection. During the summer months they hinge up to cover the collector – preventing unwanted collection, as well as protecting the cover glasses from breakage during high yard use and summer hailstorms.

EVALUATING COLLECTOR EFFICIENCY

The acid test for any solar collector is how well it performs: how many Btu's of usable heat energy per day does it put out per square foot of collector cover – after subtracting glazing supports. How well a collector performs, however, is determined by the temperature at which it is operating. The higher the temperature, the lower the efficiency.

To understand why, we must go again to collector losses. The rate of these losses is determined by a factor referred to as the "Delta T." In the case of longwave radiation losses, this is determined by subtracting the absolute temperature of outer space to the fourth power from the temperature of the collector cover glass to the fourth power. In the case of conductive losses, the "Delta T" is obtained by subtracting the outside ambient temperature (taking wind chill factors into account) from the temperature inside the collector. The higher the "Delta T," the more rapid the losses.

At some specific temperature within a given collector, a point is achieved where "stasis" occurs – that is, where losses upward and outward from the collector equal the input of heat energy. With two cover glasses, this "stasis" temperature will be about 340°F with an outside ambient temperature of 70°F for single or complex fixed-plate collectors, and somewhat higher temperature in vertical fixed-plate collector.

Perhaps a more easily understood way of explaining this is to say that the collection is the same at any temperature, but losses increase at higher temperatures. In practice, a typical fixed-plate collector will transfer twice as much sensible heat at an operating temperature of 150°F as it will at 250°F, and twice as much at 250°F as at 300°F.

Another factor affecting conductive heat losses is the action of the transfer medium itself. In hydronic systems, typically the transfer medium does not contact the sidewalls or cover of the collector. Free or natural convection does take place, however, in the area between the plate and cover, serving as a good method for transferring heat through the sidewalls and particularly through the cover and angled upper portions of the collector. (Just as in your house, where heat losses are greatest upward, requiring more insulation in the ceiling than in the sidewalls.) In forced hot-air transfer systems, on the other hand, there is forced contact between the transfer medium and the sidewalls and cover of the collector. Theoretically this should increase the losses. In practice, however, losses are not increased significantly, because boundary layers are created next to the walls and cover, which serve as a kind of insulation in operation.

As has been seen thus far, there are a number of factors commending the use of a hot-air transfer system in a solar collector. The author is certainly biased in that direction. However, in practice, it is a matter of give and take. For example, it costs more in electricity to move air than water; therefore, theoretically, a ⅛ horsepower pump can be used in a hydronic system, while a ½ horsepower motor is required for a pressurized hot air system.

However, water as a transfer medium has a number of disadvantages in terms of ordinary operation. For example, water freezes, and provisions must be made for evacuating the collector and the piping for it during winter nights when the temperature in the collector very rapidly approaches the outside ambient temperature. So some systems substitute ethylene glycol for water. But in a 5000 gallon system, about 50% might be antifreeze, or about 2500 gallons. At today's prices, that alone could bankrupt a rich man! Dust and dirt cannot be permitted in the system, and these do collect in the non-sealed system in operation. So most hydronic systems have a complicated system of settling and sedimentation filtration tanks. More bother. Additionally, even moderately hard water forms deposits on the collector which cut efficiency radically. Therefore, only distilled water or filtered rainwater can be used. Distilled water is expensive and difficult to obtain in the quantities needed, and filtered rainwater requires cumbersome collection tanks.

Taking these factors into consideration, the small increase in electrical costs (typically from $.42 per month for ⅛ horsepower to $1.67 for ½ horsepower) seems a small price indeed to pay.

Additionally, one must take into account the initial installation costs, which will be many times more for a comparable output hydronic system as compared to a hot air system. Indeed, it would probably take more than a hundred years to make up in electrical savings the additional cost of a hydronic system.

Finally, when it comes to heat storage, much higher temperatures can be achieved in hot air systems as compared to hot water systems due to the limiting factor of the boiling point of water. This means that additional savings in the cost of installation can be realized with a hot air system, since the storage area can be much smaller if a higher storage temperature can be practically achieved.

PRACTICAL MEASUREMENTS OF COLLECTOR EFFICIENCY

This section is somewhat technical and can be skipped by the casual reader.

If confronted with a given system, one of the first and most important things to check is the collector efficiency – preferably in several temperature ranges. What you want to know is how many Btu's are transferred from the collector in a given period.

You need to know the following things in order to calculate the output of the collector:

1. Temperature of the transfer medium at the inlet to the collector.
2. Temperature of the transfer medium at the outlet from the collector.
3. The flow of the transfer medium in cubic feet per minute. Most heating contractors have the equipment to measure this for you. (Note: If you are evaluating a hydronic system, you will need to convert gallons per minute to cubic feet per minute. There are 7.48 gallons of water per cubic foot.
4. The area of the collector cover after subtracting the area of glazing and glazing supports.
5. The actual time of pump or fan operation during a day's time.

It should be noted that the temperatures in 1 and 2 above may vary widely during a day's time and that mean temperatures will have to be determined. The mean temperature can then be used to evaluate the thermodynamic values for the transfer medium. These values for air and water, respectively, at standard conditions are:

Density = .075 and 62.4 lb/ft^3

Specific heat = .24 and 1.0 Btu/lb·°F

By replacing the letters in the following formula with their appropriate values for the system being evaluated, you can use ordinary arithmetic to calculate the collector output in $Btu/ft^2 \cdot day$.

$$C_o = \frac{60 \cdot F \cdot D \cdot E_t \cdot S \cdot (T_o - T_i)}{A}$$

where: C_o = Collector output, $Btu/ft^2 \cdot day$

F = Flow of transfer medium, ft^3/min

D = Density of transfer medium, lb/ft^3

E_t = Elapsed time of pump or fan operation, hr

S = Specific heat of transfer medium, Btu/lb·°F

T_o = Mean temperature of transfer medium at outlet from collector, °F

T_i = Mean temperature of transfer medium at inlet to collector, °F

A = Area of cover after subtracting area of glazing, ft^2

Sample computation for an 8 ft x 12 ft vertical fixed-plate collector located at sea level and 40° N Latitude; the collector, which uses air as the transfer medium, faces south and is inclined 60° from the horizontal; additionally, the collector is outfitted with two cover glasses and reflective surfaces of equivalent dimensions that are attached horizontally at the base of the collector:

On a sunny day in March, the mean air temperatures were measured to be 105°F at the collector inlet and 140°F at the outlet. The collection fan was found to be moving 700 cubic feet per minute at a static pressure of 2 in. of water. At the end of the day, a clock on the fan showed that it ran for 6 hours and 30 minutes. The cover area of the collector was 92 square feet after subtracting for glazing and glazing supports. Thus, by replacing the letters in the formula on the previous page with these numbers and the appropriate thermodynamic values for air (see Appendix D):

$$C_o = \frac{60 \times 700 \times .068 \times 6.5 \times .24 \times (140 - 105)}{92}$$

$$= 1695 \text{ Btu/ft}^2\text{·day}$$

Since the total daily shortwave radiation transmitted through the covers of this collector on a cloudless March day is 1939 Btu/ft^2 (see Appendix E), you can see that the collector efficiency is approximately 87%. This is partially attributable to the reflective surfaces in front of the collector which increase the percentage transmitted approximately 13%.

By not dividing by A in the equation above, you will have the total Btu collection for that day, i.e. 156,000 Btu. The "Delta T" of 35°F used in this sample computation is typical of a pressurized hot air transfer across a vertical fixed-plate collector. With a single fixed-plate collector, this "Delta T" would have been more on the order of 6 to 10°F, and the collector output less than 500 Btu/ft²·day. It is this kind of performance that is responsible for the myth that "You need a collector with one-half the square footage of the building it will heat."

THE EFFECT OF CLOUDINESS ON SOLAR COLLECTION

It often comes as a shock to a newcomer to solar energy applications that a solar collector can work on a cloudy day. The diffuse, or indirect, sky radiation is still present on a cloudy day, however, and with an efficient solar collector heat can still be captured.

For example, during periods of high, thin clouds collection may proceed at up to 80% of the normal direct rate! Even during dark, overcast days a small amount of collection may take place, particularly in the lower operational ranges if it is not raining or snowing. If precipitation is occurring, the heat is transferred very efficiently through the cover and collection does not take place.

Thus, another factor is introduced: it isn't enough to know that it's cloudy – we need to know what kind of clouds!

OTHER VARIABLES OF SOLAR COLLECTION

With fixed-plate collectors, the best collection of the day takes place during midday when the sun is at its highest and the angles of incidence are the best. Early morning and late afternoon collection are not nearly as good.

In endeavoring to predict collection, one can only use the charts prepared by the U.S. Weather Bureau (see Appendix A) which show the average percentages of cloudiness in a given month. Unfortunately, that is not sufficient information. For example, suppose that the sun shines in the early morning, then it clouds up during midday, and it clears again during late afternoon. Obviously, collection will not be as good on such a day as on one where these sunny and cloudy periods occur in the reverse order. Yet the total hours of sunshine may be the same!

The outside ambient temperatures during collection times and the wind speeds also play a part in the effectiveness of collection. So predicting collector output becomes a can of worms. The number of variables is large, and there are numbers of these for which no historical data exists. Hopefully, in years to come, as the use of solar heating increases, the Weather Bureau will begin to collect this kind of data as well.

CHAPTER 4

STORAGE OF HEAT ENERGY

Without question, the most important part of any solar heating application is the provision made for the storage of the heat energy collected during sunny periods. Any point on Earth is in darkness 50% of the time during a year. So storage must be very efficient and must be capable of containing heat energy for long periods of time. Heat is, however, difficult to store.

To understand the problems of containing heat, you must understand a few properties of heat energy. Heat is a "positive" quality; cold is a "negative" quality. Cold is the lack of heat, and we use the word "cold" in a relative or comparative way. For example, there still is heat present in an object at −200°F. Yet we would, ordinarily, say that the object was *really* cold.

In a simplistic explanation, one can equate heat with the motion of atoms – the more motion, the more heat; the less motion, the less heat.

In practical heating applications, however, "cold" is very well defined: any time the temperature in a home falls below 68°F, it's *cold!*

And anytime the temperature gets above 78°F, it is hot.

Heat is always flowing from a hotter area to a cooler area. The stronger the differential between the two areas, the more rapid the flow. This transfer is accomplished in one or more of three ways: conduction, convection, and radiation.

CONDUCTION

Practically, conduction is the flow of heat through objects. Some objects have the properties of being good heat conductors such as aluminum, while others are poor conductors such as wood. For example, if you were to insulate your home with aluminum to get the same insulating capability as you would have with just one inch of wood, you would have to use 150 feet of aluminum! To observe the conductive properties of the two materials, you could heat one end of a piece of wood while continuing to hold on to the other end, while you certainly could not do the same with a rod of aluminum without burning your hand.

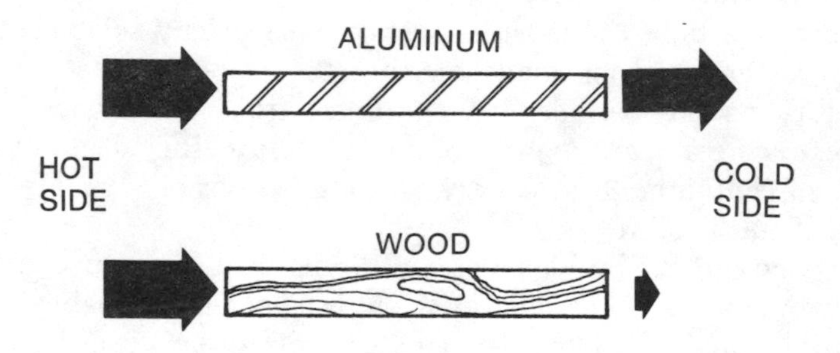

Figure 13. Conduction is dependent upon the properties of a material. Aluminum is a good conductor; heat travels rapidly through it. Wood is a poor conductor.

For conduction to occur between two objects, there must be contact between the two objects. This point of contact is sometimes referred to as a *thermal bond*. If the thermal bond between two objects is good, conduction can occur more rapidly than if the thermal bond is poor.

Heat transfer by conduction is always from hotter to cooler and may be up, down or sideways. Be sure to keep this clear in your mind, so that you don't confuse it with convection.

CONVECTION

Heat transfer by convection is accomplished by movement of air or a liquid. Hot air, for example, is less dense than cold air, and hot air rises. In any air space or liquid there is a resulting *circulation.*

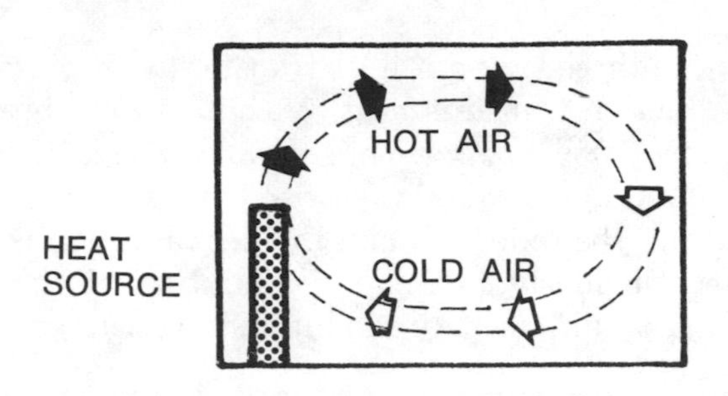

Figure 14. Hot air, being less dense, is displaced by colder air and rises.

The hot air or liquid rises and is replaced by a colder portion of the air or liquid in a continuous cycle. It is very important in storage technology to understand this process; inasmuch as the walls of a container holding air or a liquid are always being exposed to the mean temperature of the contained fluid. Stated in another way, the maximum heat contained is always being exposed to the walls and top of the container in a continuous way.

Thus, we come to a prime precept of storage technology: CONVECTION IS DETRIMENTAL TO THE STORAGE OF HEAT. From a practical standpoint, think of convection as a *heat thief*. In any system for the storage of heat where you can observe that convection is unimpeded, expect that it will not function well.

RADIATION

Actually, we have already encountered heat transfer by radiation.

Our earlier discussions of longwave and shortwave radiation in the thermal (heat) spectrum were referring to this process.

Fortunately, in storage technology, radiation is much easier to impede than conduction or convection. Why? Because most of that radiation can be *reflected* right back to the object which is radiating.

A very common-sensical and little-used device in storage areas is ordinary aluminum foil. By lining the storage area with the highly polished (reflective) foil, one can effectively impede radiative losses simply by reflecting most of the radiation right back into the storage area!

To summarize the three types of heat transfer as they affect the storing of heat: conduction transfers heat by contact, and heat flows from hot to cold and may flow up, down or sideways; convection transfers heat by circulation, and heat always rises upward; radiation is omnidirectional and may be reflected, and the net radiation is always from hotter to colder. Obviously, in a real life situation, all three processes may be occurring simultaneously and interacting with each other.

SPECIFIC HEAT STORAGE THEORY

The *specific heat* of a given substance is a term used by scientists to compare its heat acceptance capabilities as compared to water. The specific heat of water has been designated as 1.0 Btu/lb•°F. What this means is, if you have one cubic foot of water weighing 62.4 pounds and you raise its temperature by 1°F, you have "stored" 62.4 Btu's.

The specific heat of rock is about .20 Btu/lb•°F (or about 1/5 as much as that of water), so a cubic foot of rock weighing 100 pounds would only store 20 Btu's if raised 1°F.

Thus, from this *theoretical* standpoint, water is a much better storage medium than rock. In the same cubic volume of storage area, you can store 3.12 times as many Btu's for each degree rise in temperature in water as in rock. And rock costs more than water. It is due to this particular view that many "paper researchers" have specified *water* as the storage medium in their systems.

As we further examine the types of storage, we will see the pitfalls in utilizing theory instead of common sense practical research in devising storage mediums!

WATER STORAGE

As we have seen from specific heat theory, water should be an excellent storage medium. After all, it holds five times as many Btu's per degree rise in storage temperature as the same *weight* of rock and 3.12 times as many Btu's as the corresponding *volume* of rock.

One can read about many prototype systems employing huge tanks of water to store heat transferred from the collector. Often this tank of water is located in the basement of the structure being heated, so that "any heat losses from storage go right into the house or building anyhow."

A very simple practical experiment you can perform at home will allow you to determine for yourself the effectiveness of water as a storage medium. Before performing this experiment, however, you must understand one more important consideration concerning storage: it is *usable* heat that is important. A container of water at 50°F may still contain many more Btu's of heat energy than a rock which is at 120°F. When it comes to heating your home, however, a storage medium at 50°F is worse than *useless*, while the storage medium at 120°F still can provide heat for your home!

(Sometimes the author feels foolish about stating such simple and obvious common sense facts – yet many times it is just this common sense practicality which is lacking in various solar energy applications research projects going on all over the country today!)

Thus, from a practical standpoint, *usable* heat storage must be at least at 75°F or higher. Any less than this may indeed be "heat storage" from a technical viewpoint, but has no practical applicability to the heating of your home, without another heating device such as a heat pump.

Now to the experiment. Take a pint of water and a one-pound rock and heat them in your oven to the same temperature – say to 180°F. Then put them both on your kitchen counter. They both have the same amount of insulation (namely, *none)*, and no tremendous difference in surface area. Wait for one hour and come back to observe the temperatures of the two. The water is cold, while the rock is still hot!

Even if you put great quantities of insulation around each of them and repeat the experiment, you have not changed this heat loss *relationship* between the two! If you will remember our discussion of convection, you will notice that water storage has, *built-in*, the worst heat thief of all. And if you remember that each degree of temperature drop in the water releases five times as many Btu's per pound as a similar drop in the temperature of the rock, the problem is only compounded!

Thus, by applying a minor dose of common sense, one can see that water is a *terrible* storage medium, despite its theoretical superiority, due to the vastly different rates of heat loss between water and rock.

ROCK STORAGE

As we have seen in the above experiment, rock is a very good heat storage medium. It gives up its stored heat slowly, relatively speaking.

Now for a practical example of rock storage. The ISC solar furnace heat storage chamber is filled with some 25,000 pounds of granite, metamorphic forms of granite, limestone and sandstone. Lest the latter sound too technical, it is ordinary 1½ inch washed gravel!

The specific heat of this gravel is about .20 Btu/lb•°F. Using the specific heat theory, we can see that for each degree rise in temperature in the storage area, some 5000 Btu's are stored (25,000 x .20). The total usable heat storage capabilities between 75 and 225°F can be obtained by multiplying 5000 times 150° (the difference between 75 and 225), yielding a total sensible heat storage capacity of 750,000 Btu's if the storage temperature is 225°F. This is not a great quantity of heat storage if we remember that a 10,000 DD house will require about 365,000 Btu's for an average 24 hour period in January in Denver, Colorado.

HEAT OF FUSION OR "PHASE CHANGE" STORAGE

Another form of storage does not utilize the direct transfer of heat to a material which is heated and which then releases its heat as it cools. Rather, heat of fusion storage capitalizes on the fact that when a solid melts or liquifies, a certain number of Btu's are stored. When this substance resolidifies, these stored Btu's are released. A heat of fusion storage system utilizes typically eutectic salts – salts with a reasonably low melting point.

For example, a common and relatively inexpensive eutectic salt is Glauber's salt which melts at about 100°F. Approximately 108 Btu's are stored in each pound of this salt when it melts. Thus, if the storage area of a solar furnace contained 25,000 pounds of Glauber's salt, 2,700,000 Btu's could be stored when all of the salt was melted.

Unfortunately, there are a number of technical problems associated with practical utilization of heat of fusion storage. Prime among them is the requirement that the surface area of the container holding the salts be about 25 times as great as the cubic volume contained in the tank. To solve this problem, one project utilized long, narrow tubes (about the size and shape of fluorescent light tubes) to containerize the salts and still have the great amount of surface area required.

Thus, from an economic standpoint, this method of storage is quite expensive when one considers the cost of the salts and containers plus the labor of installation. Another problem is that after about 300 cycles of liquifying and resolidifying, the salts "breakdown" and must be replaced. A heating season is 270 days, and one can see a real economic problem with this method of storage! This method of storage, however, is much more technological in its approach, is more complicated, and therefore has much more appeal.

CALCULATING BTU OUTPUT FROM STORAGE

Calculating the output of a given unit is very similar to the calculations performed earlier for the collector. To set up for this, you will need to measure the flow output of the transfer medium in cubic feet per minute. Additionally, you must measure the temperature of the transfer medium at the inlet to the storage area (cold return) and at the outlet from the storage area (hot supply), the latter preferably inside the house itself. A practical formula is the following:

$$S_o = 60 \cdot F \cdot D \cdot E_t \cdot S \cdot (T_o - T_i)$$

where: S_o = Storage output, Btu

F = Flow of transfer medium, ft^3/min

D = Density of transfer medium, lb/ft^3

E_t = Elapsed time of pump or fan operation, hr

S = Specific heat of transfer medium, Btu/lb·°F

T_o = Mean temperature of transfer medium (hot supply), °F

T_i = Mean temperature of transfer medium (cold return), °F

Sample computation: You run a hot air system for ten hours, pulling heat from the storage area. Your mean temperature on the hot air supply side is 120°F, your mean cold return is 65°F. Your distribution fan delivers 1100 cubic feet per minute to the plenum inside the house. What is the Btu output per hour? By replacing the letters in the above formula with these numbers and the appropriate thermodynamic values for air (see Appendix D).

$$S_o = 60 \times 1100 \times .072 \times 10 \times .24 \times (120 - 65)$$

You obtain a total of 627,300 Btu's during the ten hour period or 62,730 Btu's per hour.

STORAGE CONFIGURATION

Below-ground storage, if an accurate cost accounting is rendered, is much more expensive than above-ground storage. Even if it is below-ground, the storage area must be properly insulated or its size radically increased to compensate for heat losses.

Excavation is expensive, even if one is planning a basement in new construction, because the storage area is not part of the usable space in that basement. After excavation is completed, expensive concrete forming and pouring must be accomplished. After that is completed, provisions must still be made for adequate insulation around that storage area.

Above-ground installations eliminate many of these costly steps, particularly if rigid polyurethane foams are used. By designing the walls so that the insulation is an integral part of the framing structure, construction costs (including materials, labor and construction profit) can be cut by as much as 50%.

Additional savings can be realized by "unitizing" the structure. For example, the ISC backyard solar furnace has a common wall of insulation between the collector and the storage area, which effectively reduces costs of insulation by about 20%.

PREVENTION OF HEAT LOSS FROM STORAGE

Since unimpeded convection is a tremendous heat thief, any properly designed storage area will have provisions for "damping" convection as much as possible. The backyard solar furnace designed by ISC, for example, has every exit and entrance to the storage area protected by a convection trap. A side view of this convection trap shown in Figure 15 will make clear the function.

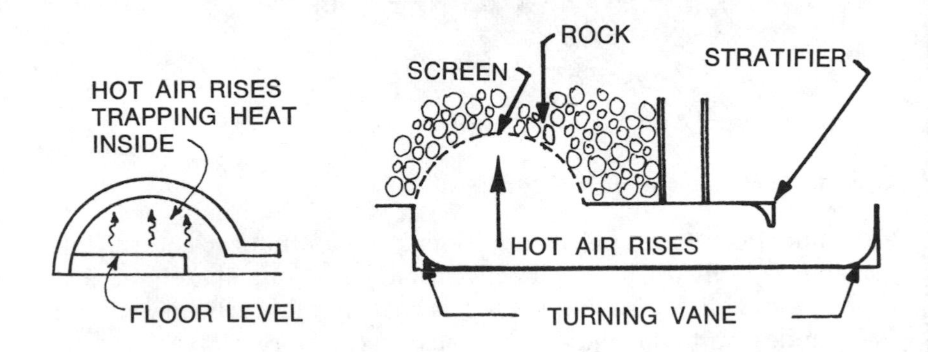

Figure 15. Convection trap. Eskimo igloo.

This common sense heat trap has been used for literally thousands of years by the Eskimos in the construction of their igloos. In free convection, *heat always rises*, so that by constructing the floor of their igloo *above* the top of the ceiling to the entrance tunnel, the hot air is contained inside the igloo living chamber itself. Obviously, convection still occurs to some degree inside the trap itself, but the escape of stored heat is drastically reduced.

In the storage area itself, minor convection still takes place in the tiny air spaces between the pieces of gravel. However, this convection tends to be contained to small areas and does not circulate in patterns throughout the chamber as a whole.

Notice also, as shown in Figure 16, that conduction is slowed as well, since the gravel touches the perimeter walls only at small points. It is only through these points that direct conduction through the side

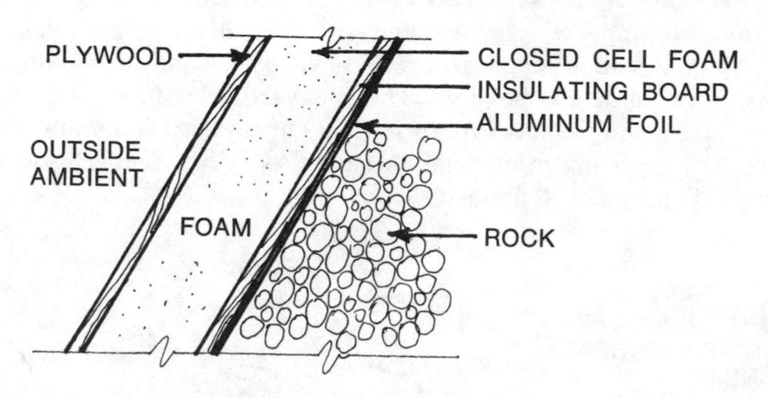

Figure 16. Configuration of storage at sidewall.

walls and floor takes place. Additional conduction, of course, takes place as a result of the small convection areas immediately adjacent to the walls. Notice also that the entire heat chamber is lined with shiny foil to reflect radiation back to the rocks in the storage area.

When examining any heat energy storage system, check to see how well it conforms to these common sense principles for containing heat. Additionally, you should check the insulating material around the storage area itself. It should have an aggregate "R Factor" of at least 25 if the storage material is rocks. See Appendix D for thermal property data for various insulating materials.

CHAPTER 5

THE ISC BACKYARD SOLAR FURNACE

In the preceding text, many references have been made to the ISC backyard solar furnace auxiliary home heating system, comparing it to other types of solar heating devices. Obviously, from the author's viewpoint, as one of the inventors of the unit, it is a paradigm against which to measure other systems. The reader can judge for himself whether or not this is true.

Certainly, from a historical standpoint, this system deserves recognition as the first factory-produced solar furnace and as the first working system compact enough to be compatible to existing homes. If for no other reasons than these, the system deserves a comprehensive examination.

An outside view of the 96 square foot collector model is shown in Figure 17. This unit, when installed, measures 8 feet 11¼ inches by 12 feet 8⅜ inches at the base and stands just over eight feet high at the peak of the triangle. The reflective shields measure 9 feet by 12 feet and hinge up in summertime to cover the solar collector. The outside of the unit is finished in rough sawn fir, stained an earthtone brown.

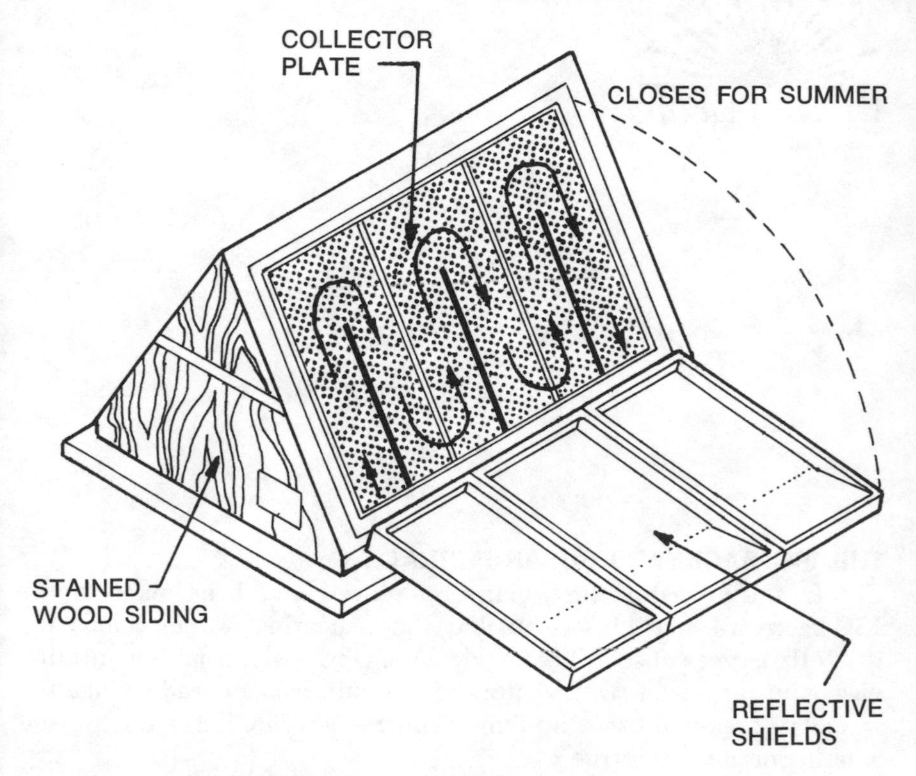

Figure 17. Outside view of the ISC backyard solar furnace.

As can be seen in the floor plan of the unit shown in Figure 18, there are two fans in the unit. One is utilized for collection, the other for distribution of hot air to the home being heated. These fans operate independently and may be operating at the same time if house heat is required during the day while the collector fan is operating.

As can be seen in Figure 19, the collector fan pulls air through the heat storage area and down through a convection trap at the fan entrance. As the air leaves the fan, it is pushed downward through another convection trap and thence upward into the collector. The air

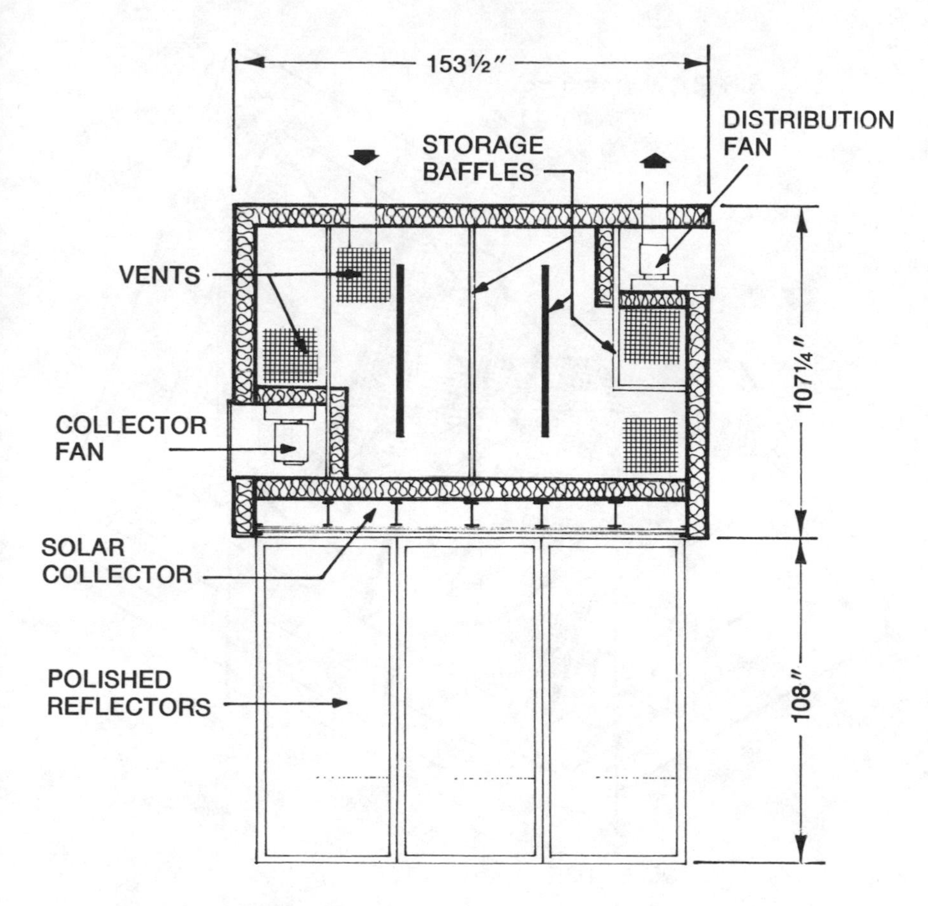

Figure 18. Floor plan of the ISC backyard solar furnace.

is then moved up and down in an S-shaped path across the collector until it exits downward on the other side of the collector through still another convection trap, thence upward through a screen into the gravel storage area. Notice that the hot air traveling through the heat storage area is required by virtue of baffles placed vertically in its path to traverse an S-shaped path through the rock storage as well, returning through another convection trap to the collector fan.

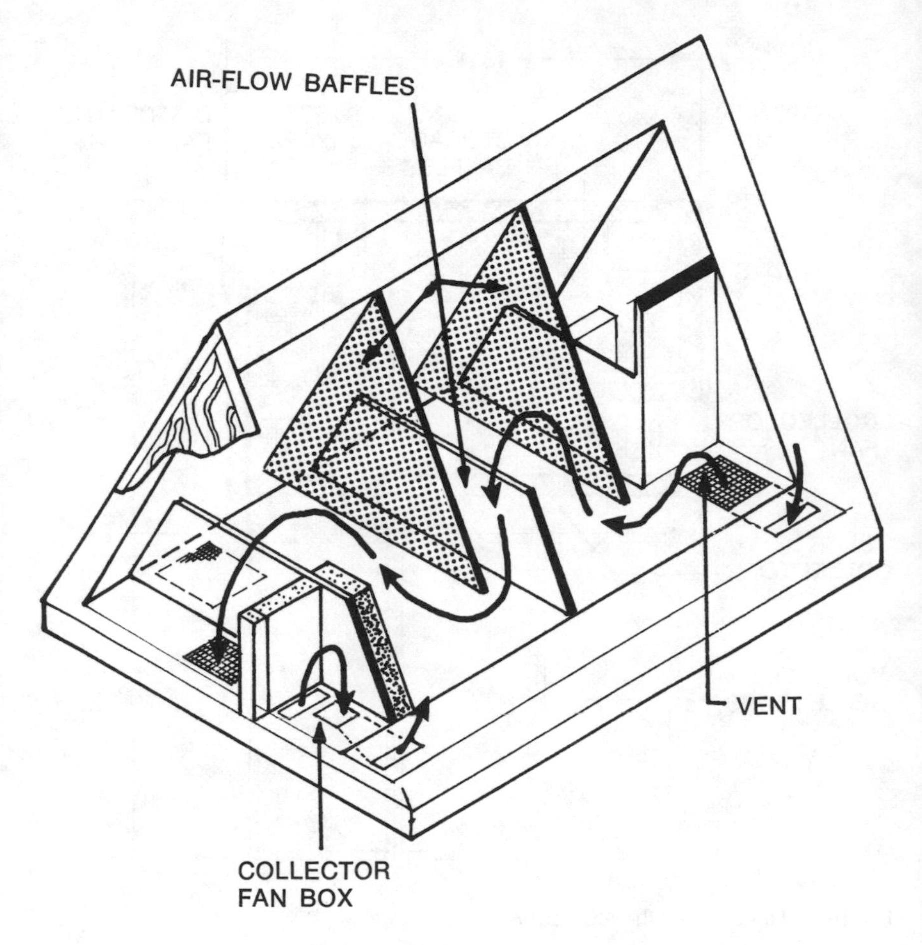

Figure 19. Cut away drawing of the ISC backyard solar furnace.

It is very important to the function of the unit that the air passing over the face of the collector be pressurized sufficiently to set up a turbulence around the openings of the aluminum vanes extending upward into the air stream, since it is this turbulence which effectively "washes" the heat accumulated in the cups into the air stream for transference to the storage area.

CONSTRUCTION

The backyard furnace is self-foundational, utilizing a 10 by 14 foot polystyrene foam base pad 6 inches in thickness and protected by a plastic vapor barrier. This pad is set on leveled, undisturbed ground, and requires no excavation other than the minimal amount required to create a level spot. It extends outward from the unit 6 inches in every direction to insulate the ground around the unit, so that the formation of ice lenses and concomitant frost heave will not shift the unit itself.

The floor of the storage chamber is also of 6-inch-thick polystyrene foam laminated between plywood and a special insulating washed wood fiber cross-laminated board. All walls are built from 4-inch-thick polyurethane rigid foam laminated board and rough sawn fir plywood siding. All interior surfaces of the storage area and interior ducting are lined with shiny foil.

All framing supports are 20 or 24 gauge galvanized steel, as are the collector glass supports. Glass panels measure 4 feet by 8 feet, are 3/16 inches thick, and are double glazed. The entire unit is hermetically sealed with specially formulated high temperature acrylic caulking and high temperature polyurethane foam.

Fans are industrial quality heavy duty with angle iron reinforcing and sealed ball bearings, specially designed to function well under high static pressure loads. These are belt-driven by ½ horsepower thermally protected quality motors.

Hidden wire-ways are provided for all internal circuitry. All framing components are prefabricated in factories in 4 foot by 8 foot or 9 foot modules which can be easily assembled on site by authorized distributors or do-it-yourselfers. The washed ¾ to 1½ inch gravel is loaded into the unit on site, as well.

CONTROLS

The collection fan is controlled by sophisticated solid state comparator circuitry. A thermistor probe in the collector and another in the storage area feed the existing temperatures at the two points into a logic circuit which compares the two and allows the fan to operate only when the temperature in the collector exceeds that in the storage area by a predetermined number of degrees. This control is extremely important, because it prevents cooling of the stored heat on cloudy days by preventing unnecessary cycling of the fan. Additionally, it prevents breakage of the under cover glass due to thermal shock.

Also important is the fact that switching operations are not performed by relays — notorious for malfunctioning due to burnt contacts. Instead, switching is performed by solid state triacs, in this case rated to carry a continuous load of 25 amperes and surges up to 1000 amperes. (Each of the ½ horsepower motors draws approximately 7½ amperes with starting surges up to 40 amperes).As a further protection against damage to controls, all circuity is embedded in a solid block of epoxy, preventing both mechanical and moisture damage.

Another function built into the collection fan control is an automatic resetting device. If, for example, a power outage should occur during the midday time on a bright sunny day, when the power was restored, the fan could easily cause breakage of the cover glasses due to the extremely large temperature differential which might exist between the collector plate and the inlet air stream to the collector. After a power outage, the automatic resetting device cycles the fan on and off rapidly for a period of time to gradually reduce this temperature differential before resuming regular collection functions.

All high voltage circuitry (115 volt) is separated from the low voltage circuitry (24 volt) and is connected optically, so that accidental shock hazards are reduced.

The house distribution fan is controlled in two ways. The prime control is your existing house thermostat. Whenever the temperature in your house drops below the setting of your house thermostat, the distribution fan is turned on to supply hot air to your house.

The hot supply air to your home is pushed from the solar furnace downward through a convection trap and thence through connecting ducting (which may be above or below ground) to the plenum of your present forced-air furnace. Cold return air is pulled from your home through another connecting duct upward through a convection trap into the rock storage chamber to be heated again.

Referring back to Figure 19, you will see that the direction of this air flow is just the reverse of the collection air flow through the storage area. This is due to the fact that there is a gradation of temperatures, typically, in the battery. The "hot" side of storage is the portion nearest the collector outlet, the "cool" side is nearest the collector fan. By reversing the flow for the house supply, the cold air returning to the unit from the house is introduced on the "cool" side of the storage area.

What happens if extended sunless periods cause the battery to "run down," that is, cause the rock to drop below an operable temperature? Then both furnaces operate in tandem. When the battery drops to 75°F, your present furnace takes over completely, so that your home heating continues uninterrupted.

As soon as the heat storage is sufficiently "recharged," this control switches off your present furnace, and the solar furnace takes over again. All automatically!

INSTALLATION

On existing homes, the solar furnace is installed in the yard as close as is feasible to the house, with a due south orientation. The collector must not be shaded by surrounding obstructions at any time during the day. Electrical lines supplying the unit are preferably installed underground, but may be installed overhead to a mast on the unit. Ducting connecting the unit to the house may be above or below ground, at the homeowner's option.

In new construction, several options are possible. First, of course, the unit can be installed as an auxiliary building, just as on existing homes. Another possibility is that of installing the unit as part of the new structure, by building it right into a south wall of the house. This leaves primarily just the collector extending outside the house, somewhat like a bay window. With this type of installation, a 4 foot by 12 foot closet is formed inside the house, part of which can be used to house an in-line electrical duct heater or a conventional forced-air fueled furnace as the standby unit. The remaining space can be utilized as a pantry or as a clothes or storage closet.

Finally, if the builder so decides, the unit can be installed on some portion of the roof of the structure – on the flat roof of a garage for example. Such installations are more expensive, since engineering the extremely high weight loads is necessary. One design advantage to a roof installation is that when the reflective shields are closed for the summer the resulting available roof space can provide an attractive upper sun deck for outdoor use.

CARE OF A SOLAR FURNACE

A question that is often asked pertains to keeping the collector clean. In operation, dust and smog collect on the outer surface of the collector. Surprisingly, even rather heavy deposits of dirt on the collector face do not interfere drastically with collection. It has been observed that a loss of only about 4 to 8% in collection efficiency results. Fortunately, Mother Nature tends to keep the collector face cleansed automatically, with occasional rains and snows.

Another question pertains to snow build-ups on the face of the collector in wintertime. Obviously, if the snow is swept off immediately, more collection time is gained. Fortunately, even if nothing is done, the sun will relatively quickly solve the problem. Due to the steep angle of the collector face, the slickness of the glass and heat losses upward from the collector, the snow slides to the bottom of the collector where additional collector losses melt it fairly quickly. Snow removal from the reflectors is not necessary, since the snow itself is a pretty good reflector!

Glass breakage is another concern. Obviously, if a neighborhood menace throws a rock at the collector, breakage will occur. Just as if it were your picture window that were the target! The glass is a heavy-duty, 3/16-inch-thick glass – much heavier than that used typically for windows in residences. Of course, the solar furnace should be added to your homeowner's insurance policy.

The reflective shields protect this glass during the summer months when the kids are out of school and when yard use is at its highest. Hail storms, another potential hazard, primarily occur in summertime also, so that the doors protect against damage there too.

Once a year, the blowers and motors should be serviced and the belts checked. As with a conventional furnace, the air filters should be changed three times during the heating season. For convenience, the heating contractor should install these duct filters inside the house, thereby making accessibility easier.

CHAPTER 6

WHY AN AIR SYSTEM CAN BE TEN TIMES SMALLER THAN A WATER SYSTEM

The major problem with solar research in the past has been that it has been specific, as opposed to approaching the problem from a total systems standpoint. An excellent collection system is worse than useless if it is hooked up to a bad storage system. The various parts of the system must be *matched* to each other! You would not buy a $500 speaker system to hook into a $29.95 stereo, yet many solar heating systems do just this sort of thing.

Let us examine the least expensive water systems, the all-hydronic type. Many of the problems which will be pointed out do have possible solutions, but they increase the cost and complexity of the system. An all-hydronic system has a roof-top collector and an insulated water tank for storage in the basement. Heat from storage is distributed throughout the house through baseboard radiators. Such a system has the advantage of being compatible to existing hot water heating systems.

The first problem arises with collector efficiency. A flat-plate collector operating at 200°F will function at no more than 30% efficiency, while that same collector will function at a 90 to 95% efficiency at 100°F. (Collector efficiency is simply the amount of energy transferred to storage divided by the amount of solar energy that gets inside the collector.) Pure common sense says that if you operate the system in the 100°F range you're better off. If you operate in the 200°F range, you must have roughly *three times* as much collector to do the same job as you would need if you operated in the more efficient 100°F range.

But the baseboard radiators will not function at much less than 160°F. In fact, the fueled hot water system generally works in the 185 to 200°F range. This means that the collector must always work in the 200°F range in an all-hydronic system, if it is to collect *useful heat!*

In the all-air system, however, proper sizing of the system to the house can keep the collector operating in the 100°F range during the critical midwinter months when heating demands are highest. Typically, the all-air system will have a downpoint temperature of 75°F. (Downpoint is the temperature at which the solar furnace shuts off and the fueled furnace takes over totally).

If this were the only problem with an all-hydronic system, it would have to be only three times larger to do the same job, and would only cost three times more. But an even bigger problem exists. Even with good insulation, it is very difficult to keep a water tank above 160°F. For example, you may have had experience with trying to keep coffee hot for a long period in a thermos bottle. A thermos bottle has excellent insulating qualities; better, in fact, than the insulation used on water tanks. Yet after just 24 hours, that coffee has cooled way below 160°F.

This cooling of the storage tank below 160°F presents a terrific problem when the system is operating under real world conditions.

It is not unusual for three days of cloudiness in a row to occur during the midwinter months, and it is here that the problem arises. After 24 hours, the system has cooled to the downpoint of 160°F or lower. After 72 hours, you can be quite sure that the temperature will be approaching 90°F if the tank is in the basement with a room temperature of 70°F. Even if the sun comes out on the fourth day, the system must collect *non-useful* energy to get back up to the downpoint temperature of 160°F before it can start collecting useful energy. Even with a relatively small water storage tank, the recovery required can be staggering. For example, a small 1500 gallon water tank will require 840,000 Btu's to raise the temperature from 90°F back up to 160°F. This means having about 750 to 1000 square feet of collector working *all one sunny day!*

Thus, some 1000 to 1300 square feet of collector are required in the all-hydronic system, to deliver upon demand the same amount of heat to the residence as a properly engineered 100 square feet of air collector with pebble bed storage!

The difference in cost is appreciable. Any system, air or water, will cost you, the consumer, from $25.00 to $40.00 per square foot of collector for the total system installed on your house. This means the difference between $4000 and at least $25,000. Unfortunately, not all manufacturers point out this little $21,000 difference in cost—rather, the system is broken up into components. "The collector only costs $7.50 per square foot F.O.B. the factory." You can find out later that the whole system, installed, with such a collector price, will cost $25,000 to $40,000! A good economic rule of thumb is to multiply the factory price of a *collector* by four to obtain the finished *installed* price of the *total system.*

AVERAGE CASE ENGINEERING

In most areas of the United States, you should install a system capable of supplying 75 to 90% of your heat, NOT MORE. To go from 90 to 100% solar heating requires at least quadrupling the size of the system, meaning that you are buying at least three more systems just to handle that last 10%, which is simply not economical.

Therefore, your solar heating system is, and should be, an *auxiliary* heating system! It may sound strange to say that a system supplying the home with 90% of its heating needs is an auxiliary, but it should be just that, for economic reasons. The fueled or electric forced-air furnace should be sized to supply all the worst case conditions, because the one thing predictable about the weather is that it is not predictable. This means that sometimes the solar furnace is going to be depleted, so you must have 100% backup for it. This being the case, the solar engineer should take full advantage of the main furnace and design your solar furnace to handle some percentage of *average* heating requirements.

For example, a well-insulated 1000 square foot home will require about 375,000 Btu's per day at 10°F below zero. The main furnace should then be of a size to handle this requirement. The *average* requirement of heat for that house, however, will only be about 115,000 Btu's per day during the heating season in Denver, Colorado and only 145,000 Btu's per day in Green Bay, Wisconsin. Thus, the size of the system required is considerably reduced by providing some percentage of the average requirements as opposed to the worst case requirements.

As a matter of fact, this is where good old-fashioned common sense comes into play. If you design the size of the solar heating system to provide 75% of the energy required in January, you will have a system which is producing 700% of what is required in September in most areas. So that you must trade off the over-production during 6 months against the under-production during 3 months to obtain the most economical system. In most parts of the country, an overall supply of 75 to 90% of the heat required during the heating season is an economic amount. Programming the solar furnace to reach downpoint, or to "runout" fairly regularly (about 25% of the time) during December, January and February has the advantage of keeping the collector operating temperatures in the lower temperature ranges, hence, at the highest efficiency during the highest demand heating months. Such engineering is mostly common sense, but requires a careful eye be kept on costs.

One widely used "rule of thumb" which is making the rounds presently is one you should beware of. It goes like this: "You need a collector half the square footage of the home." This, despite its wide circulation, is totally asinine! Does this rule apply in Atlanta, Georgia or Juneau, Alaska? The *only* way a solar furnace can be sized to a home is by performing a heat loss calculation! You will be able to do this yourself by following the instructions in the next chapter. But even with that information, no rule of thumb of collector size is possible, since the size of the collector is dependent both on the sun conditions and the winter weather conditions in your locale.

The reader is strongly advised to follow the simple precautions set forth in Appendix G when preparing to purchase a solar heating system. You may save *literally* tens of thousands of dollars by doing so!

CHAPTER 7

THE PRACTICALITY OF HOME SOLAR HEATING

Whether one is considering the practicality of solar heating for an existing home or a home to be built, common sense tests must be applied to the situation.

Foremost among these is *solar exposure*. After all, sunshine is the fuel for a solar furnace, and if a given location by reason of topography or surrounding structures or trees does not have direct access to sunshine, it is obviously foolish to install a solar furnace.

To check this out is very simple. Look at the situation. If you have a 40-story high-rise condominium south of your home, or a huge flowering shade tree to the south of the proposed location for your furnace, solar heating is not practical. The collector must not be shaded by anything during the day.

If you live on the north side of an east-west street, the unit must be located in the side yard or far enough back in your back yard not to be shaded by your home during the low-sun-angle winter months. If you have a typical one-story home, the unit must be at least 46 feet behind the house; if two-story, at least 64 feet behind the house. Probably the unit should not be located any further than 75 feet away from the house due to heat losses during transmission from the unit to your home.

HEAT LOSS CALCULATIONS

The next important consideration is the heating requirements of your home. These must be calculated from the size of your home, the insulation it has, the size and number of windows and doors, etc.

A relatively simple, if less than totally accurate method for performing heat loss calculations on a given residence follows. You will need to have the following data to perform these calculations:

1. Measurement of the perimeter of your house and the ceiling height.
2. The sizes of each of the doors and windows in your house.
3. The square feet of ceiling below the attic, and the square feet of the bottom floor (if you have no basement).

Taking these measurements, you then convert all of them into square feet measurements. For example, let's assume that the perimeter of your house is 180 lineal feet (45 feet by 45 feet by 45 feet by 45 feet) and your ceilings are 8 feet high. This means that you have a total of 1440 square feet of outside walls. Suppose that your house has 15 windows and 3 doors, and adding up the total square footage, you have 150 square feet of window area and 60 square feet of door area. Finally, assume that your house is a ranch style home with 2025 square feet on the main floor and has a full basement. You have 2 inches of sidewall insulation, 3⅝ inches of ceiling insulation, single pane windows and no storm sash.

Using the heat loss estimating multipliers for a design temperature of —10°F from the chart on the next page, you can determine that the sidewall heat losses will be 14,400 Btu/hr (1440 x 10); window losses, 13,500 Btu/hr (150 x 90); door losses, 2340 Btu/hr (60 x 39); ceiling losses, 14,175 Btu/hr (2025 x 7). The floor losses will be 0. Adding these losses, we find that for this typically insulated 2025 square-foot home, the heat losses at design temperature will be a total of 44,415 Btu/hr. To see the difference a little more insulation can make, perform the same calculations for a house having the same dimensions, but with 4 inches of sidewall insulation, 10 inches of ceiling insulation and storm windows and doors. The hourly heat loss is reduced to 22,525 Btu/hr or approximately one-half as much!

HEAT LOSS ESTIMATING MULTIPLIERS

	Design Outside Temperature (°F)						
	−30	−20	−10	0	+10	+20	+30
Sidewalls: Frame or Brick*							
No insulation	26	23	21	18	16	13	10
2″ insulation	13	12	10	9	8	7	5
3⅝″ insulation	7	6	6	5	4	4	3
Windows:							
Single glazing	113	102	90	79	68	57	45
Double glazing	69	62	55	48	41	35	28
Doors:							
1½″ solid wood	49	44	39	34	29	25	20
1½″ solid wood with metal storm door	33	30	26	23	20	17	13
Ceilings: Attic above*							
No insulation	32	29	26	22	19	16	13
3⅝″ insulation	8	7	7	6	5	4	3
6″ insulation	5	5	4	4	3	3	2
10″ insulation	3	3	2	2	2	2	1
14″ insulation	2	2	2	2	1	1	1
Floors:							
Concrete slab	10	9	8	7	6	5	4
Over crawl space (no insul)	14	13	11	10	8	7	6
Over crawl space (4″ insul)	5	5	4	4	3	3	2
Over heated basement	0	0	0	0	0	0	0

*Based on fiberglass batt insulation ($k = 0.3$ Btu/ft²•hr•°F/in.).

CONVERSION TO DEGREE DAYS

Heat loss calculations are performed for extreme temperature conditions and are utilized to obtain the size heating plant required to heat a house under severe winter conditions. In Denver, for example, with a design temperature of —10°F, the number of days each winter when the temperature actually drops to −10°F is extremely small.

Indeed, studies of temperatures made over long periods of time show that the average mean temperature in Denver in January is +28°F!

Every part of the construction of a solar furnace is expensive. Constructing the unit to perform totally on the few days when temperatures do drop to very low levels is simply uneconomical – particularly when there is a standby fueled furnace. In fact, enlarging the unit to handle those few times each winter could easily quadruple the cost of the solar furnace.

Therefore, a solar heating system which is practical will be designed to provide the heat required, or some percentage thereof, during *average* heating conditions during the winter months.

To estimate these average heating requirements, we use the "Degree Day." Degree Days (DD) are figured from a base of 65°F outside ambient temperature. They are averages of temperatures taken in a given locality over a long period of years. A table of Degree Days for various cities is presented in Appendix A. It is assumed with the Degree Day method that if the outside temperature is 65°F or higher, no heating requirements exist to keep the inside temperature of the home at 70°F. Each degree below a 65°F outside temperature is a Degree Day. For example, if in a given month the average mean outside temperature is 55°F, then each day has 10 Degree Days.

Using the Degree Day method, one can estimate very closely the number of Btu's required to heat a given structure in a given locality for a given month. And although it is an *averaging* system, it is extremely practical. As a result, it has widespread use by heating contractors everywhere.

After you have calculated the design temperature requirements for your home, it is a simple matter to convert that Btu figure into a figure usable with the Degree Day method.

Say that your house has a calculated heat loss of 30,060 Btu/hr at a design temperature of −10°F. In a 24 hour period at that temperature, the house will require 721,400 Btu's (30,060 x 24) to maintain an inside temperature of 70°F. Between the base 65°F and −10°F, there are 75 Degree Days. Dividing the 24 hour requirement of 721,400 Btu's by 75, we obtain a requirement of 9620 Btu's for each Degree Day. We would therefore designate this house as a 9620 Btu/Degree Day house. This means that if the outside temperature were 64°F, it would require 9620 Btu's to heat the home for 24 hours.

A simpler method for doing this conversion exists, however, by using the chart below. You can multiply the Btu loss per hour you calculated for your home by the appropriate conversion factor listed, and you will obtain the size of your home in Btu/Degree Days.

CONVERSION FACTORS FOR HOURLY BTU HEAT LOSS TO BTU/DD SIZE

	Design Outside Temperature (°F)						
	−30	−20	−10	0	+10	+20	+30
Conversion Factor	.25	.28	.32	.37	.44	.53	.69

USING THE DEGREE DAY TABLES TO CALCULATE HEATING REQUIREMENTS

It may be useful to perform a sample calculation of Btu requirements utilizing the Degree Day method. Referring to Appendix A, we find that in Denver the number of Degree Days in January is 1132. We can multiply the size of a home expressed in Btu/Degree Days and determine the average heating requirements. For example, using our 9620 Btu/DD house, we multiply 9620 times 1132, obtaining 10.89×10^6 Btu's of heat energy required to heat this house during the entire month of January. To obtain the *average* daily requirement, we divide by 31, obtaining 351,300 Btu's per day. Similar calculations can be performed for each of the winter months using the Degree Day table.

HOW WELL WILL THE SOLAR FURNACE PERFORM?

As we have seen, there are a tremendous number of variables which affect the performance of a solar furnace. Solar insolation is at different levels at different times of the year. Its angles with relationship to the solar furnace change continuously during the day and from day to day. Cloudiness must be accounted for – not only the percentage of cloudiness, but the type of cloudiness and when during the day it occurs. Wind and outside temperatures have an effect on its performance as well. The location of the furnace also is important — one part of a community may be shaded by clouds while another part of the community has sunshine on a given occasion. Finally, the installation itself has tremendous variables. Two identical houses can have vastly different fuel bills—one may be kept at 75°F, the other at 68°F–one may have windows open in wintertime, the other may not. Then other variables enter the picture: how well is the house insulated, how good is the weatherstripping, etc. Heat loss calculations tell you nothing until the energy consciousness of the family living in the house, as well as its living habits, are taken into account.

Therefore, estimates of efficiency become little more than guesses. However, one needs to know about what to expect in terms of efficiency. Therefore, a good practice is to have a professional heating contractor look over the situation in person to see just what the installation itself will be like. Although he will still be "guessing" there will be more real data involved in his guess!

You may want to structure an estimate on a given solar heating system yourself. A reasonable method for doing this is to calculate the Btu/DD size of the structure to be heated, and using the Degree Day tables, list the month by month requirements to heat that structure. Then, using output data calculated by yourself on the collector involved (see Chapter Three for practical methodology), list the month by month collection capability of the collector as if every day were a full sun day. Then, using the table from Appendix A that lists mean percentage of possible sunshine, reduce the collection figures by multiplying them times the appropriate percentage. If you only have a month or so of data, the chart in Appendix A that shows approximate cloudless day solar insolation may be helpful in extrapolating for the months where you have no data.

Obviously, such a method is only approximate due to the many variables listed earlier, but it may help you to evaluate a given system.

ENERGY CONSERVATION

Whether or not you decide to heat your home with solar energy, energy conservation is needed *now!* By some figures, home heating accounts for 25% of all the energy consumed in the United States. By effecting some of the measures listed below, you can make a significant contribution to easing the energy crisis and a contribution to your pocketbook as well.

It has been stated by experts that 30% more heat is required to maintain your home at 76°F than at 70°F. Therefore, it makes very good sense to keep your thermostat setting as low as is comfortable, since tremendous energy savings can be effected by this one action.

A number of things reduce your heating load by producing heat. For example, one 100-watt light bulb burning for 24 hours produces nearly 8200 Btu's. Seated, watching television, the average adult will add 390 Btu's to the home each hour. So the size of your family can actually have an effect on your heating bill!

Your house itself, for example, is a kind of solar furnace if you let it be. Windows on the south side of your home permit entry of much heat energy, as you probably know. But windows on the north side of your house do too. If you have a window on the north side of your house that is only 10 square feet in size, on a typical cloudless day in December, it will permit entry of more than 2000 Btu's. A south window the same size will permit entry of about 20,000 Btu's per day! Therefore, it makes tremendous sense to open all the drapes in your house during the daytime in winter months and close them at night. This simple expedient can make a major contribution to reducing your heating load.

The vent fan above your range, if it is vented to the outside, is a villain when it comes to conserving heat. Modify it, if possible, to vent to the inside of the house through a charcoal filter. If that is not possible, use the vent in a sparing way during winter months. Your fireplace is another heat thief. Add glass doors to it. This will still permit total enjoyment of the fireplace and will pay for itself very quickly in heat savings by cutting the amount of chimney draft.

If you have an electric dryer, it is a good practice to vent it to the inside of the house in winter months. Again, tremendous amounts of heat energy you are paying for are blown right outside! You get an additional advantage from this in that the low winter humidity levels in your house are raised, at least temporarily, which will make the house more comfortable at a lower temperature. (CAUTION: Do *not* vent a gas dryer to the inside of the house — you'll end up gassing yourself!)

Most important of all is insulation. The effect of proper insulation simply can not be stressed enough. As we saw in an earlier example, heating loads were doubled by virtue of just a *little* less insulation. INSULATION IS CHEAP! One of the biggest mistakes made universally by insulation salesmen is in trying to amortize the cost of insulation against the fuel savings. This is ridiculous. The additional insulation you put in your home becomes part of your real estate EQUITY. You will be able to get more money for your house when you sell it if it has adequate insulation as opposed to poor insulation. It's just common sense — like any other home improvement, it adds to the value of your house. It is an investment, therefore, not an expense. You get it back. So any fuel savings are bonus dividends on that investment.

Next weekend, add some dollars to the equity in your home. Take the time to add two 6-inch batts of fiberglass insulation to your attic. The cost is minimal, the labor a matter of climbing into the attic and laying the batts between the rafters.

If you don't have or can't afford storm windows, a very inexpensive substitute which works nearly as well is to tape ordinary plastic film on the insides of your windows, a quarter inch or more out from the glass. You can get big sheets of this film at any hardware store.

Whether or not you opt for solar heating, these simple suggestions can make a world of difference in your heating bills!

CHAPTER 8

IS SOLAR HEATING PRACTICAL NOW?

Residential solar heating is practical now in most of the United States. To determine if your home is suitable for solar heating, however, you must use a common sense, practical approach.

First, you must look at the location of your home and yard with respect to the sun. Sunshine is the fuel for your solar furnace, and unless you can get adequate sunshine, you are going to be disappointed in the performance of your unit. So be very realistic in appraising the practicality of solar heating for your home by seeing if you have a functional location for a solar furnace. If you don't have a good place to locate a collector, don't let your enthusiasm for solar heating lead you into a minimal installation.

Second, run heat loss calculations on your home. See just what the Btu/DD size of your home is. If it exceeds the values shown in the chart on the next page, you should add sufficient insulation and storm sashes to make your Btu/DD size consistent with your square footage.

TYPICAL BTU/DD SIZES OF WELL INSULATED HOMES

Square Feet*	Btu/DD Size
750	4000
1000	5000
2000	7500
3000	10000
4000	12500
5000	15000

*Areas do not include heated basements.

Third, check the Btu/DD recommendations of the manufacturer of the unit you intend to install.

Fourth, using a chart prepared by the manufacturer of estimated efficiency in your city, see that the percentage of annual needs the unit will supply are satisfactory for you.

ECONOMIC CONSIDERATIONS OF SOLAR HEATING

Fossil fuels are being depleted at an alarming rate. Therefore, the prices for these fuels are escalating almost daily. In the next few years, the price of fuel to heat a home in January very well may exceed the mortgage payment for that month! Obviously, no one can afford that kind of ratio.

More importantly, the day is coming when rationing of heating fuels will become severe – to the point that one will not be able to keep his home warm! In Britain, for example, this is already happening – where one room in the house is kept warm on a part-time basis.

Solar heating is a practical answer to these problems – not to mention the considerations of personal comfort and health. No price tag can or should be put on the latter.

THE FIRST LAW OF PRACTICAL SOLAR ENERGY UTILIZATION

"To be practical, any solar energy device (residential or commercial) must be compatible to existing buildings." Why? Because there are 65 million existing homes and countless existing commercial buildings. And only a few new buildings (by comparison) are being built each year.

Even if *every* new building in the United States were equipped with solar heating, it would take 40 years to get just 50% utilization of solar energy! And the sad fact of the matter is – we don't have 40 years, and we must effect more than 50% utilization of solar energy. Our fossil fuel reserves are being depleted so rapidly that only a crash program to equip *all existing and new buildings* with solar heating will have any appreciable effect on the energy crisis. Therefore, any practical system must be compatible to existing structures.

The backyard solar furnace will require some getting used to. It is not the most aesthetic structure in the world, but if aesthetics were grounded on function and purpose, the backyard solar furnace would be the most beautiful art form in man's history!

APPENDIX A
CLIMATOLOGICAL DATA*

CONTENTS

*Reproduced from the *Climatic Atlas of the United States*, Environmental Data Service, U.S. Department of Commerce, June 1968.

NORMAL TOTAL HEATING DEGREE DAYS, JANUARY
(Base 65°)
NOTE.--CAUTION SHOULD BE USED IN INTERPOLATING ON THESE GENERALIZED MAPS, PARTICULARLY IN MOUNTAINOUS AREAS.
THESE MAPS ARE BASED ON THE 30-YEAR PERIOD, 1931-60
SCALE 1:10,000,000
ALBERS EQUAL AREA PROJECTION — STANDARD PARALLELS 29½° AND 45½°
ALASKA
HAWAII
PUERTO RICO AND VIRGIN ISLANDS
GULF OF MEXICO
ATLANTIC OCEAN
PACIFIC OCEAN

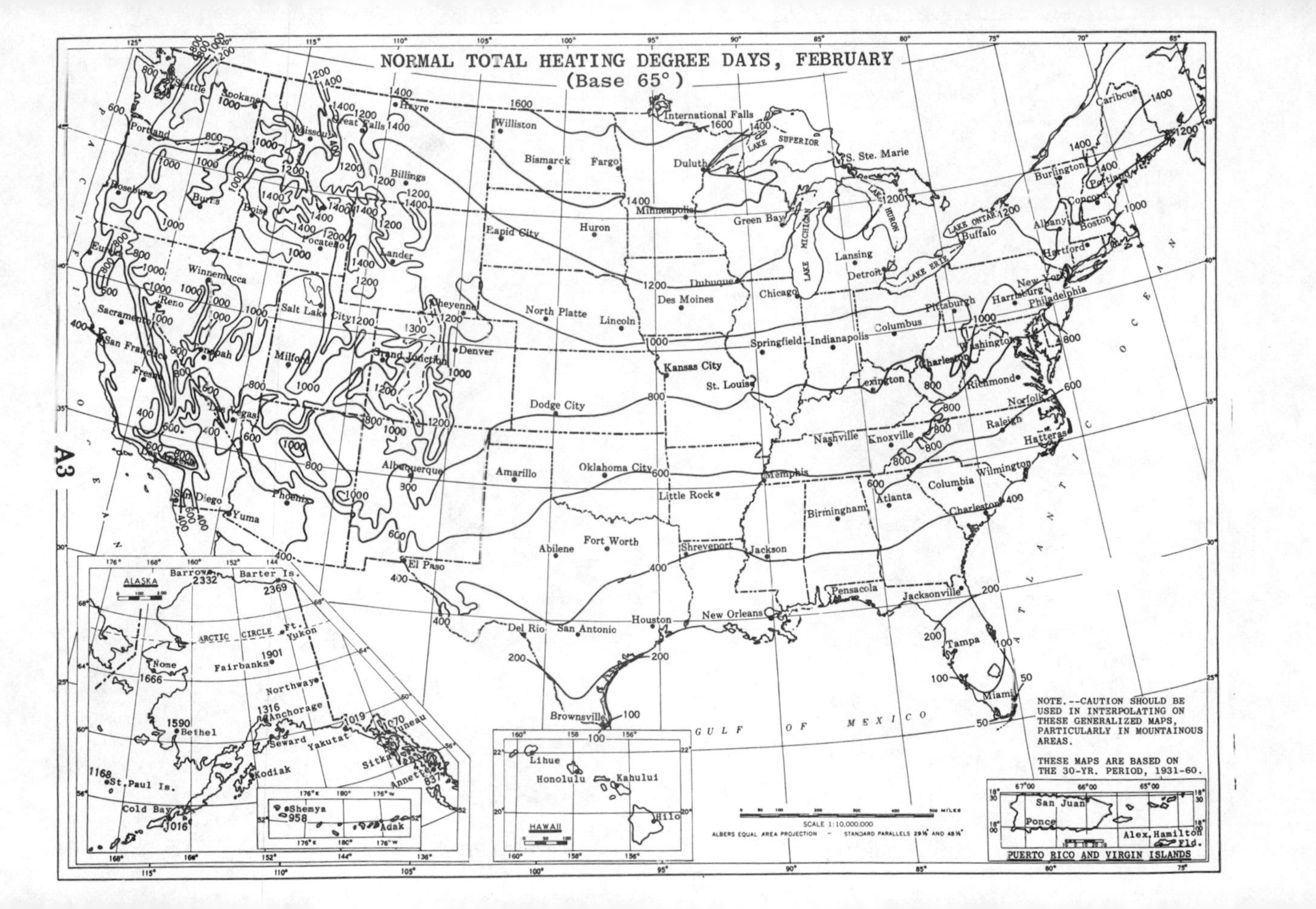

NORMAL TOTAL HEATING DEGREE DAYS, FEBRUARY
(Base 65°)
NOTE.--CAUTION SHOULD BE USED IN INTERPOLATING ON THESE GENERALIZED MAPS, PARTICULARLY IN MOUNTAINOUS AREAS.
THESE MAPS ARE BASED ON THE 30-YR. PERIOD, 1931-60.
SCALE 1:10,000,000
ALBERS EQUAL AREA PROJECTION — STANDARD PARALLELS 29½° AND 45½°
GULF OF MEXICO
ATLANTIC OCEAN
PACIFIC OCEAN
ALASKA
HAWAII
PUERTO RICO AND VIRGIN ISLANDS
Seattle
Spokane 1000
Portland
Roseburg
Eureka
Burns
Boise
Pendleton
Missoula
Great Falls
Havre
Billings
Pocatello
Lander
Winnemucca
Reno
Sacramento
San Francisco
Fresno
Tonopah
Salt Lake City
Milford
Las Vegas
San Diego
Yuma
Phoenix
Cheyenne
Denver
Grand Junction
Albuquerque
El Paso
Williston
Bismarck
Fargo
Rapid City
Huron
North Platte
Lincoln
Dodge City
Amarillo
Oklahoma City
Abilene
Fort Worth
Del Rio
San Antonio
Houston
Brownsville
International Falls
Duluth
Minneapolis
Des Moines
Dubuque
Kansas City
St. Louis
Little Rock
Shreveport
New Orleans
Jackson
Memphis
Green Bay
Chicago
Springfield
Indianapolis
Lansing
Detroit
S. Ste. Marie
Columbus
Lexington
Nashville
Knoxville
Birmingham
Atlanta
Pensacola
Jacksonville
Tampa
Miami
Pittsburgh
Charleston
Washington
Richmond
Norfolk
Raleigh
Hatteras
Wilmington
Columbia
Buffalo
Albany
Hartford
New York
Harrisburg
Philadelphia
Boston
Concord
Burlington
Portland
Caribou
LAKE SUPERIOR
LAKE MICHIGAN
LAKE HURON
LAKE ERIE
LAKE ONTARIO
Barrow 2332
Barter Is. 2369
ARCTIC CIRCLE
Ft. Yukon
Fairbanks 1901
Nome 1666
Northway
Anchorage 1316
Bethel 1590
Seward
Yakutat 1019
Juneau 1070
Sitka
Annette 837
Kodiak
St. Paul Is. 1168
Cold Bay 1016
Shemya 958
Adak
Lihue
Honolulu
Kahului
Hilo
San Juan
Ponce
Alex. Hamilton Fld.

A3

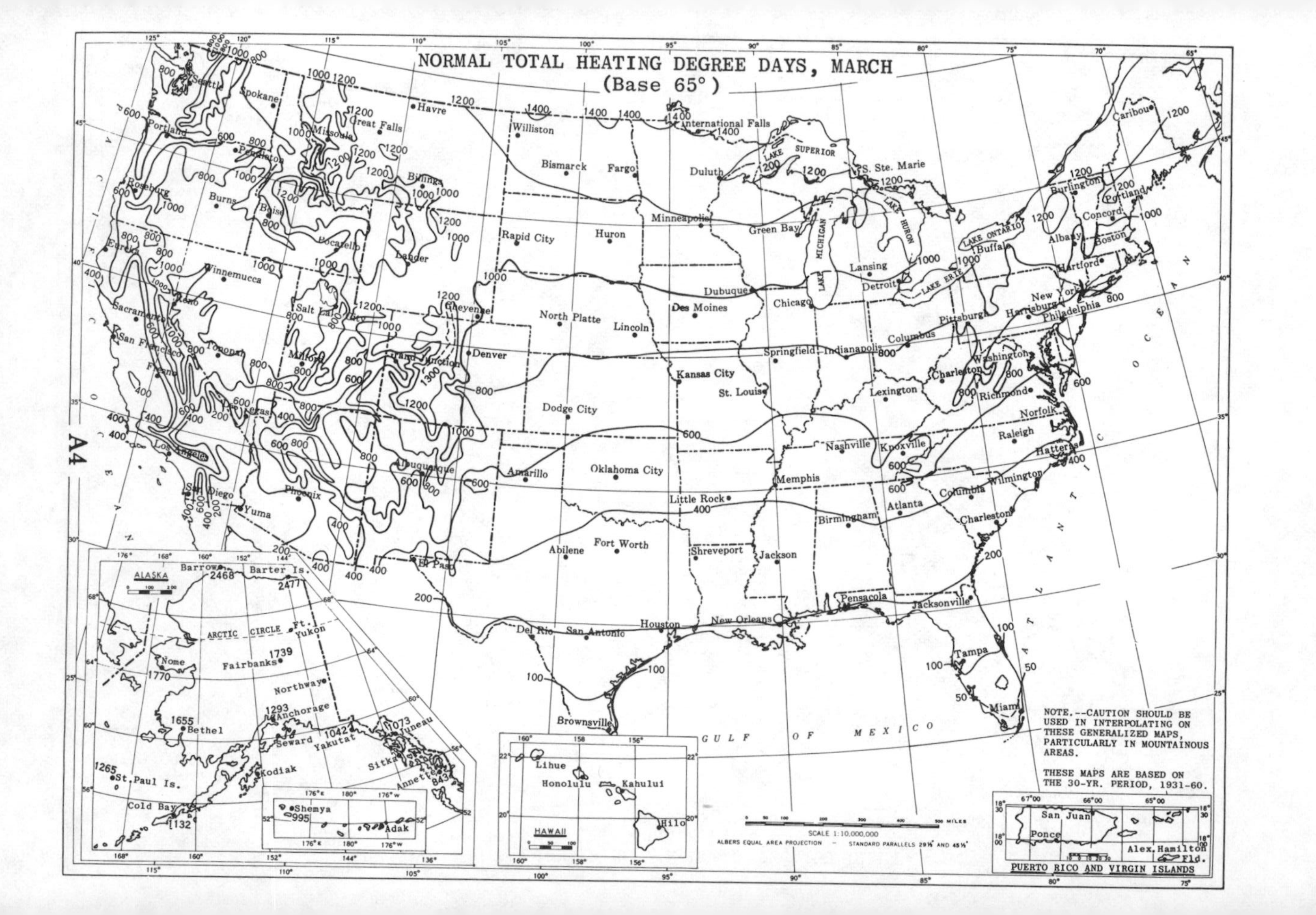
NORMAL TOTAL HEATING DEGREE DAYS, MARCH
(Base 65°)
NOTE.--CAUTION SHOULD BE USED IN INTERPOLATING ON THESE GENERALIZED MAPS, PARTICULARLY IN MOUNTAINOUS AREAS.
THESE MAPS ARE BASED ON THE 30-YR. PERIOD, 1931-60.
SCALE 1:10,000,000
ALBERS EQUAL AREA PROJECTION — STANDARD PARALLELS 29½° AND 45½°
GULF OF MEXICO
ALASKA
Barrow 2468
Barter Is. 2477
Ft. Yukon
Fairbanks 1739
Nome 1770
Northway
Anchorage 1293
Bethel 1655
Seward
Yakutat 1042
Juneau 1073
Sitka
Annette 843
Kodiak
St. Paul Is. 1265
Cold Bay 1132
Shemya 995
Adak
HAWAII
Lihue
Honolulu
Kahului
Hilo
PUERTO RICO AND VIRGIN ISLANDS
San Juan
Ponce
Alex. Hamilton Fld.

NORMAL TOTAL HEATING DEGREE DAYS, APRIL
(Base 65°)
NOTE.--CAUTION SHOULD BE USED IN INTERPOLATING ON THESE GENERALIZED MAPS, PARTICULARLY IN MOUNTAINOUS AREAS.
THESE MAPS ARE BASED ON THE 30-YR. PERIOD, 1931-60.
SCALE 1:10,000,000
ALBERS EQUAL AREA PROJECTION — STANDARD PARALLELS 29½° AND 45½°
ALASKA
HAWAII
PUERTO RICO AND VIRGIN ISLANDS
GULF OF MEXICO
ATLANTIC OCEAN
PACIFIC OCEAN
Havre
Williston
Bismarck
Fargo
International Falls
Duluth
S. Ste. Marie
Caribou
Spokane
Great Falls
Missoula
Portland
Pendleton
Billings
Burlington
Roseburg
Burns
Boise
Minneapolis
Green Bay
Concord
Albany
Boston
Huron
Pocatello
Lander
Lansing
Hartford
Eureka
Winnemucca
Detroit
Dubuque
Chicago
New York
Reno
Salt Lake City
Cheyenne
North Platte
Des Moines
Pittsburgh
Harrisburg
Philadelphia
Sacramento
San Francisco
Tonopah
Milford
Grand Junction
Denver
Lincoln
Springfield
Indianapolis
Columbus
Washington
Fresno
Kansas City
St. Louis
Lexington
Charleston
Richmond
Dodge City
Norfolk
Las Vegas
Raleigh
Nashville
Knoxville
Hatteras
Los Angeles
Albuquerque
Amarillo
Oklahoma City
Memphis
Wilmington
San Diego
Phoenix
Little Rock
Atlanta
Columbia
Yuma
Birmingham
Abilene
Fort Worth
Shreveport
Jackson
El Paso
Pensacola
Jacksonville
Del Rio
San Antonio
Houston
New Orleans
Tampa
Miami
Brownsville
LAKE SUPERIOR
LAKE MICHIGAN
LAKE HURON
LAKE ONTARIO
LAKE ERIE
Barrow 1944
Barter Is. 1923
ARCTIC CIRCLE
Ft. Yukon
Fairbanks 1068
Nome 1314
Northway
Anchorage 879
Bethel 1173
Seward
Yakutat 840
Juneau 810
Sitka
Annette 663
St. Paul Is. 1098
Kodiak
Cold Bay 951
Shemya 885
Adak
Lihue
Honolulu
Kahului
Hilo
San Juan
Ponce
Alex. Hamilton Fld.

NORMAL TOTAL HEATING DEGREE DAYS, MAY
(Base 65°)
NOTE.--CAUTION SHOULD BE USED IN INTERPOLATING ON THESE GENERALIZED MAPS, PARTICULARLY IN MOUNTAINOUS AREAS.
THESE MAPS ARE BASED ON THE THE 30-YR. PERIOD, 1931-60.
SCALE 1:10,000,000
ALBERS EQUAL AREA PROJECTION — STANDARD PARALLELS 29½° AND 45½°
ALASKA
HAWAII
PUERTO RICO AND VIRGIN ISLANDS
GULF OF MEXICO
PACIFIC OCEAN
ATLANTIC OCEAN
LAKE SUPERIOR
LAKE MICHIGAN
LAKE HURON
LAKE ONTARIO
LAKE ERIE
ARCTIC CIRCLE
Barrow 1445
Barter Is. 1373
Ft. Yukon
Fairbanks 555
Nome 930
Northway
Anchorage 592
Bethel 806
Seward
Yakutat 632
Juneau 601
Sitka
Annette 505
Kodiak
St. Paul Is. 936
Cold Bay 791
Shemya 837
Adak
Lihue
Honolulu
Kahului
Hilo
San Juan
Ponce
Alex. Hamilton Fld.

NORMAL TOTAL HEATING DEGREE DAYS, JUNE
(Base 65°)
NOTE.--CAUTION SHOULD BE USED IN INTERPOLATING ON THESE GENERALIZED MAPS, PARTICULARLY IN MOUNTAINOUS AREAS.
THESE MAPS ARE BASED ON THE 30-YR. PERIOD, 1931-60.
SCALE 1:10,000,000
ALBERS EQUAL AREA PROJECTION — STANDARD PARALLELS 29½° AND 45½°
ALASKA
HAWAII
PUERTO RICO AND VIRGIN ISLANDS
GULF OF MEXICO
PACIFIC OCEAN
ATLANTIC OCEAN

NORMAL TOTAL HEATING DEGREE DAYS, JULY
(Base 65°)
NOTE.--CAUTION SHOULD BE USED IN INTERPOLATING ON THESE GENERALIZED MAPS, PARTICULARLY IN MOUNTAINOUS AREAS.
THESE MAPS ARE BASED ON THE 30-YR. PERIOD, 1931-60.
SCALE 1:10,000,000
ALBERS EQUAL AREA PROJECTION — STANDARD PARALLELS 29½° AND 45½°
GULF OF MEXICO
ALASKA
HAWAII
PUERTO RICO AND VIRGIN ISLANDS

NORMAL TOTAL HEATING DEGREE DAYS, AUGUST
(Base 65°)
NOTE.--CAUTION SHOULD BE USED IN INTERPOLATING ON THESE GENERALIZED MAPS, PARTICULARLY IN MOUNTAINOUS AREAS.
THESE MAPS ARE BASED ON THE 30-YR. PERIOD, 1931-60.
SCALE 1:10,000,000
ALBERS EQUAL AREA PROJECTION — STANDARD PARALLELS 29½° AND 45½°
ALASKA
Barrow 840
Barter Is. 775
ARCTIC CIRCLE
Ft. Yukon
Fairbanks 332
Nome 496
Northway
Anchorage 291
Bethel 394
Seward
Yakutat 347
Juneau 338
Sitka
Annette 223
Kodiak
St. Paul Is. 539
Cold Bay 425
Shemya 475
Adak
HAWAII
Lihue
Honolulu
Kahului
Hilo
PUERTO RICO AND VIRGIN ISLANDS
San Juan
Ponce
Alex. Hamilton Fld.
GULF OF MEXICO
ATLANTIC OCEAN
PACIFIC OCEAN
LAKE SUPERIOR
LAKE MICHIGAN
LAKE HURON
LAKE ERIE
LAKE ONTARIO
Seattle
Spokane
Portland
Pendleton
Roseburg
Burns
Boise
Eureka
Winnemucca
Reno
Sacramento
San Francisco
Tonopah
Fresno
Las Vegas
Los Angeles
San Diego
Yuma
Phoenix
Missoula
Great Falls
Havre
Billings
Pocatello
Lander
Salt Lake City
Milford
Cheyenne
Grand Junction
Denver
Albuquerque
El Paso
Williston
Bismarck
Fargo
Rapid City
Huron
North Platte
Lincoln
Dodge City
Amarillo
Oklahoma City
Abilene
Fort Worth
Del Rio
San Antonio
Houston
Brownsville
International Falls
Duluth
Minneapolis
Green Bay
S. Ste. Marie
Dubuque
Des Moines
Kansas City
St. Louis
Springfield
Chicago
Indianapolis
Lansing
Detroit
Columbus
Lexington
Little Rock
Shreveport
Memphis
Nashville
Knoxville
Jackson
Birmingham
Atlanta
New Orleans
Pensacola
Jacksonville
Tampa
Miami
Columbia
Charleston
Wilmington
Raleigh
Hatteras
Norfolk
Richmond
Washington
Pittsburgh
Harrisburg
Philadelphia
New York
Buffalo
Albany
Hartford
Boston
Concord
Portland
Burlington
Caribou

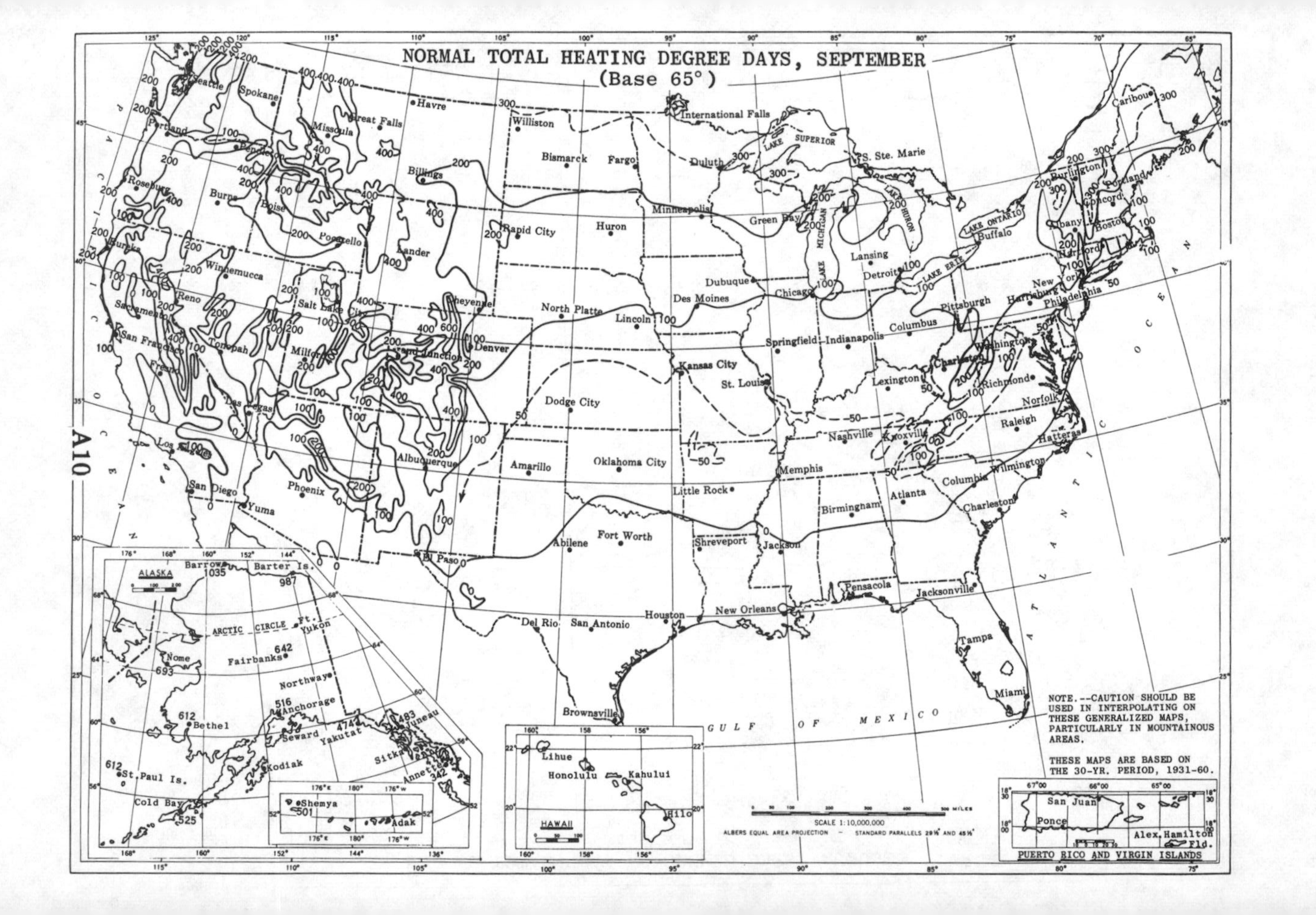
NORMAL TOTAL HEATING DEGREE DAYS, SEPTEMBER
(Base 65°)
NOTE.--CAUTION SHOULD BE USED IN INTERPOLATING ON THESE GENERALIZED MAPS, PARTICULARLY IN MOUNTAINOUS AREAS.
THESE MAPS ARE BASED ON THE 30-YR. PERIOD, 1931-60.
SCALE 1:10,000,000
ALBERS EQUAL AREA PROJECTION — STANDARD PARALLELS 29½° AND 45½°
ALASKA
HAWAII
PUERTO RICO AND VIRGIN ISLANDS
GULF OF MEXICO
ATLANTIC OCEAN
PACIFIC OCEAN

NORMAL TOTAL HEATING DEGREE DAYS, OCTOBER
(Base 65°)
NOTE.--CAUTION SHOULD BE USED IN INTERPOLATING ON THESE GENERALIZED MAPS, PARTICULARLY IN MOUNTAINOUS AREAS.
THESE MAPS ARE BASED ON THE 30-YR. PERIOD, 1931-60.
SCALE 1:10,000,000
ALBERS EQUAL AREA PROJECTION — STANDARD PARALLELS 29½° AND 45½°
PACIFIC OCEAN
ATLANTIC OCEAN
GULF OF MEXICO
ALASKA
HAWAII
PUERTO RICO AND VIRGIN ISLANDS
Seattle
Spokane
Portland
Pendleton
Roseburg
Burns
Boise
Eureka
Winnemucca
Reno
Sacramento
San Francisco
Tonopah
Fresno
San Diego
Yuma
Phoenix
Milford
Salt Lake City
Pocatello
Havre
Great Falls
Missoula
Billings
Lander
Cheyenne
Denver
Grand Junction
Albuquerque
El Paso
Williston
Bismarck
Fargo
Rapid City
Huron
North Platte
Lincoln
Dodge City
Amarillo
Oklahoma City
Abilene
Fort Worth
Del Rio
San Antonio
Houston
Brownsville
International Falls
Duluth
Minneapolis
Des Moines
Dubuque
Kansas City
St. Louis
Little Rock
Shreveport
Memphis
Jackson
New Orleans
Green Bay
Chicago
Springfield
Indianapolis
Lansing
Detroit
S. Ste. Marie
Columbus
Lexington
Nashville
Knoxville
Birmingham
Atlanta
Pensacola
Jacksonville
Tampa
Miami
Pittsburgh
Charleston
Washington
Richmond
Norfolk
Raleigh
Hatteras
Wilmington
Columbia
Charleston
Harrisburg
Philadelphia
New York
Hartford
Albany
Boston
Buffalo
Burlington
Concord
Portland
Caribou
LAKE SUPERIOR
LAKE MICHIGAN
LAKE HURON
LAKE ERIE
LAKE ONTARIO
Barrow 1500
Barter Is. 1482
ARCTIC CIRCLE
Ft. Yukon
Fairbanks 1203
Nome 1094
Northway
Anchorage 930
Bethel 1042
Seward
Yakutat 716
Juneau 725
Sitka
Annette 567
Kodiak
St. Paul Is. 856
Cold Bay 772
Shemya 784
Adak
Lihue
Honolulu
Kahului
Hilo
San Juan
Ponce
Alex. Hamilton Fld.

NORMAL TOTAL HEATING DEGREE DAYS, NOVEMBER
(Base 65°)
NOTE.--CAUTION SHOULD BE USED IN INTERPOLATING ON THESE GENERALIZED MAPS, PARTICULARLY IN MOUNTAINOUS AREAS.
THESE MAPS ARE BASED ON THE 30-YR. PERIOD, 1931-60.
SCALE 1:10,000,000
ALBERS EQUAL AREA PROJECTION — STANDARD PARALLELS 29½° AND 45½°
PACIFIC OCEAN
ATLANTIC OCEAN
GULF OF MEXICO
ALASKA
HAWAII
PUERTO RICO AND VIRGIN ISLANDS
Seattle
Spokane
Havre
Williston
International Falls
Duluth
Bismarck
Fargo
Minneapolis
Green Bay
S. Ste. Marie
Caribou
Burlington
Portland
Concord
Albany
Boston
Hartford
New York
Philadelphia
Harrisburg
Pittsburgh
Buffalo
Detroit
Lansing
Chicago
Dubuque
Des Moines
Lincoln
North Platte
Rapid City
Huron
Billings
Great Falls
Missoula
Pendleton
Portland
Roseburg
Burns
Boise
Pocatello
Lander
Cheyenne
Denver
Grand Junction
Salt Lake City
Milford
Winnemucca
Reno
Eureka
Sacramento
San Francisco
Fresno
Tonopah
Las Vegas
Los Angeles
San Diego
Yuma
Phoenix
Albuquerque
El Paso
Amarillo
Dodge City
Oklahoma City
Abilene
Fort Worth
Del Rio
San Antonio
Houston
Brownsville
Shreveport
Little Rock
Kansas City
St. Louis
Springfield
Indianapolis
Columbus
Lexington
Charleston
Washington
Richmond
Norfolk
Raleigh
Hatteras
Wilmington
Columbia
Charleston
Atlanta
Knoxville
Nashville
Memphis
Birmingham
Jackson
New Orleans
Pensacola
Jacksonville
Tampa
Miami
LAKE SUPERIOR
LAKE MICHIGAN
LAKE HURON
LAKE ONTARIO
LAKE ERIE
Barrow 1971
Barter Is. 1944
ARCTIC CIRCLE
Ft. Yukon
Fairbanks 1833
Nome 1455
Northway
Anchorage 1284
Bethel 1434
Seward
Yakutat 936
Juneau 921
Sitka
Annette 738
Kodiak
St. Paul Is. 963
Cold Bay 921
Shemya 876
Adak
Lihue
Honolulu
Kahului
Hilo
San Juan
Ponce
Alex. Hamilton Fld.

NORMAL TOTAL HEATING DEGREE DAYS, DECEMBER
(Base 65°)

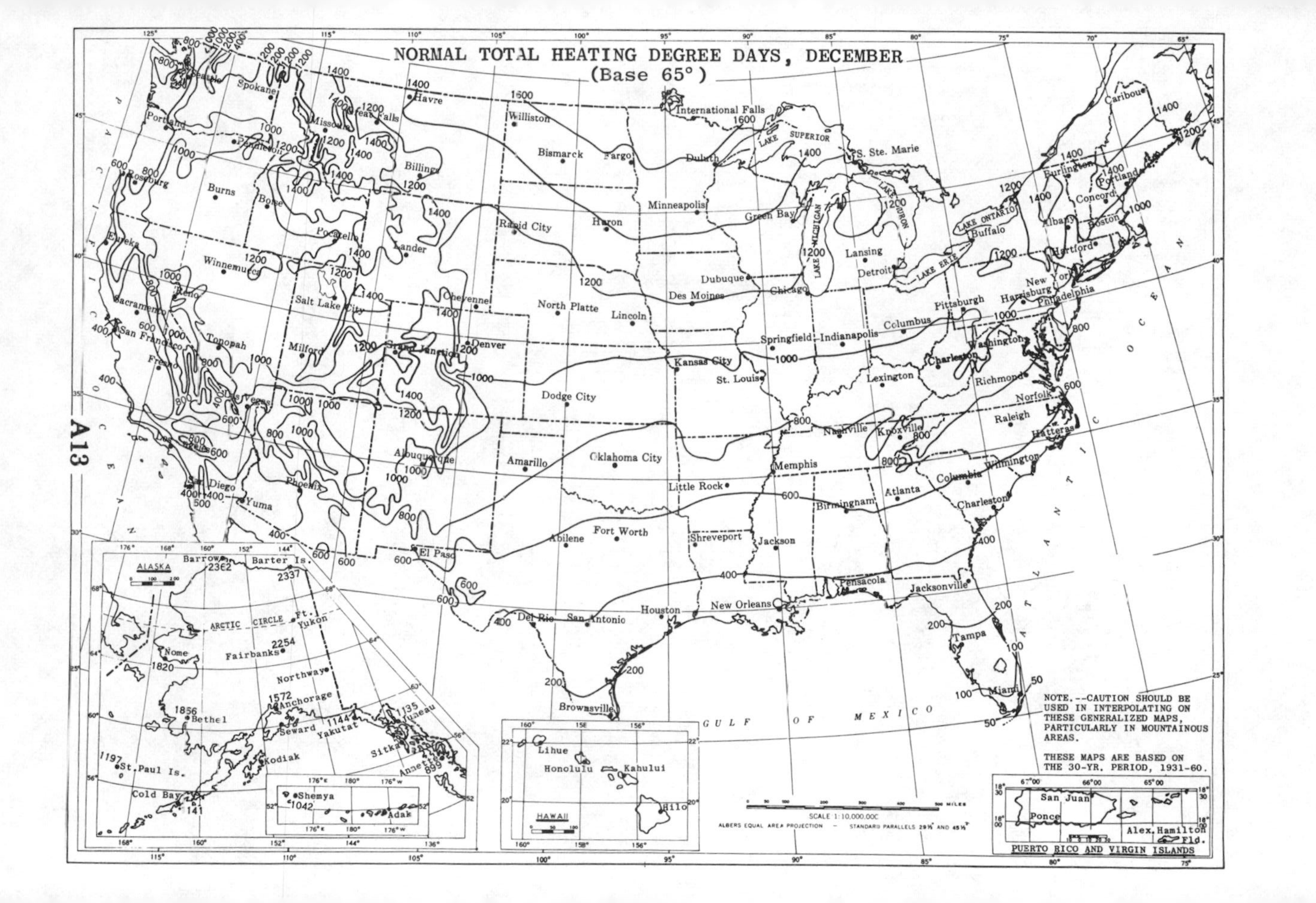

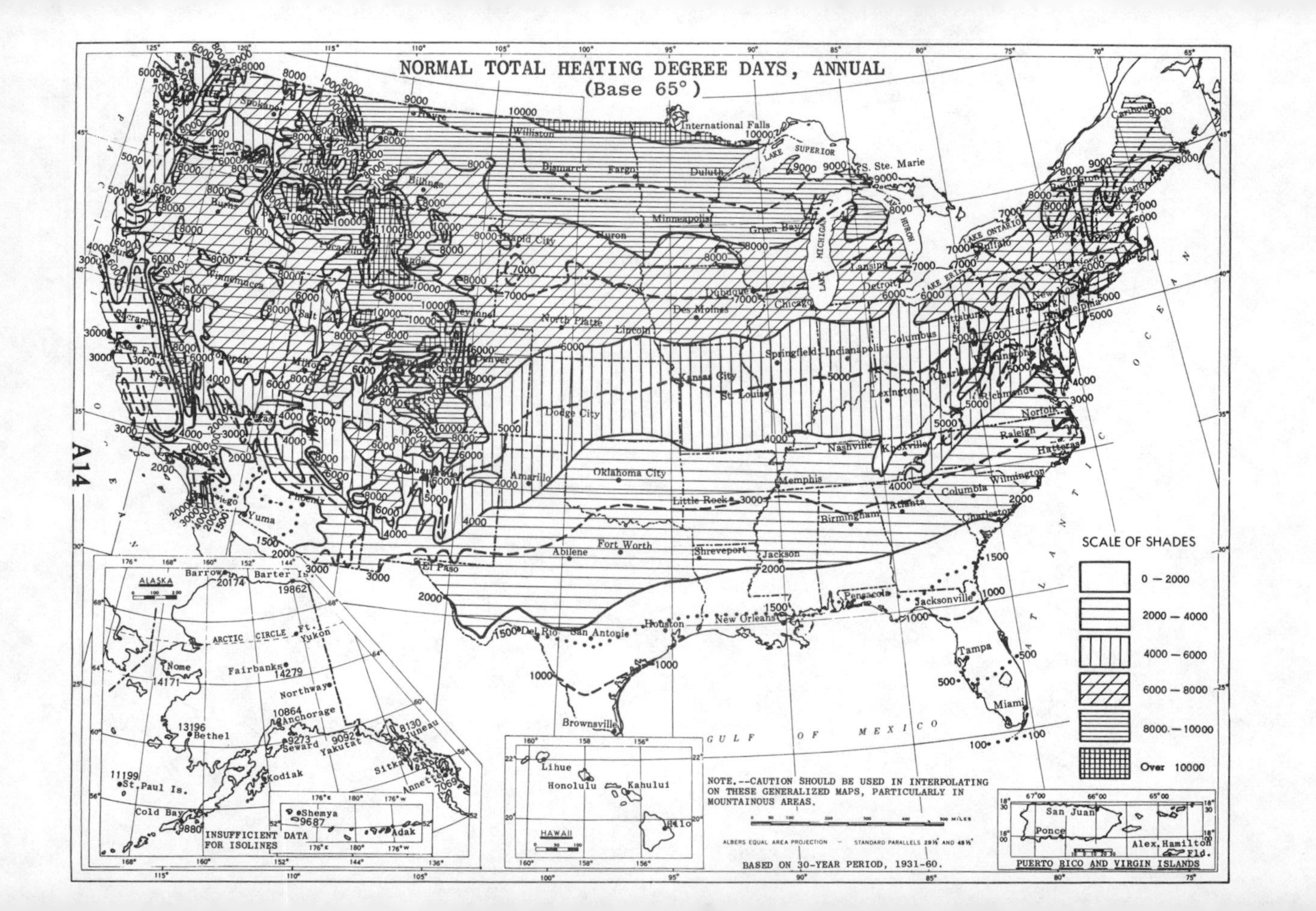
NORMAL TOTAL HEATING DEGREE DAYS, ANNUAL
(Base 65°)
SCALE OF SHADES
0 — 2000
2000 — 4000
4000 — 6000
6000 — 8000
8000 — 10000
Over 10000
NOTE.--CAUTION SHOULD BE USED IN INTERPOLATING ON THESE GENERALIZED MAPS, PARTICULARLY IN MOUNTAINOUS AREAS.
ALBERS EQUAL AREA PROJECTION — STANDARD PARALLELS 29½ AND 45½
BASED ON 30-YEAR PERIOD, 1931-60.
GULF OF MEXICO
ATLANTIC OCEAN
PACIFIC OCEAN
ALASKA
INSUFFICIENT DATA FOR ISOLINES
HAWAII
PUERTO RICO AND VIRGIN ISLANDS
International Falls
Williston
Bismarck
Fargo
Duluth
Minneapolis
Green Bay
Huron
Rapid City
North Platte
Lincoln
Des Moines
Dubuque
Chicago
Lansing
Detroit
Springfield
Indianapolis
Columbus
Pittsburgh
Kansas City
St. Louis
Lexington
Dodge City
Denver
Cheyenne
Billings
Spokane
Burns
Winnemucca
Salt
Milford
Phoenix
Yuma
Amarillo
Oklahoma City
Little Rock
Memphis
Nashville
Knoxville
Birmingham
Atlanta
Columbia
Charleston
Wilmington
Raleigh
Hatteras
Norfolk
Richmond
Harrisburg
Buffalo
Caribou
Abilene
Fort Worth
Shreveport
Jackson
El Paso
Del Rio
San Antonio
Houston
New Orleans
Pensacola
Jacksonville
Tampa
Miami
Brownsville
S. Ste. Marie
LAKE SUPERIOR
LAKE MICHIGAN
LAKE HURON
LAKE ERIE
LAKE ONTARIO
Barrow 20174
Barter Is. 19862
ARCTIC CIRCLE
Ft. Yukon
Nome 14171
Fairbanks 14279
Northway
10864 Anchorage
13196 Bethel
9273 Seward
9092 Yakutat
8130 Juneau
Sitka
Annette 7069
11199 St. Paul Is.
Kodiak
Cold Bay 9880
Shemya 9687
Adak
Lihue
Honolulu
Kahului
Hilo
San Juan
Ponce
Alex. Hamilton Fld.

NORMAL TOTAL HEATING DEGREE DAYS (Base 65°)

STATE AND STATION	JULY	AUG.	SEP.	OCT.	NOV.	DEC.	JAN.	FEB.	MAR.	APR.	MAY	JUNE	ANNUAL
ALA. BIRMINGHAM	0	0	6	93	363	555	592	462	363	108	9	0	2551
HUNTSVILLE	0	0	12	127	426	663	694	557	434	138	19	0	3070
MOBILE	0	0	0	22	213	357	415	300	211	42	0	0	1560
MONTGOMERY	0	0	0	68	330	527	543	417	316	90	0	0	2291
ALASKA ANCHORAGE	245	291	516	930	1284	1572	1631	1316	1293	879	592	315	10864
ANNETTE	242	208	327	567	738	899	949	837	843	648	490	321	7069
BARROW	803	840	1035	1500	1971	2362	2517	2332	2468	1944	1445	957	20174
BARTER IS.	735	775	987	1482	1944	2337	2536	2369	2477	1923	1373	924	19862
BETHEL	319	394	612	1042	1434	1866	1903	1590	1655	1173	806	402	13196
COLD BAY	474	425	525	772	918	1122	1153	1036	1122	951	791	591	9880
CORDOVA	366	391	522	781	1017	1221	1299	1086	1113	864	660	444	9764
FAIRBANKS	171	332	642	1203	1833	2254	2359	1901	1739	1068	555	222	14279
JUNEAU	301	338	483	725	921	1135	1237	1070	1073	810	601	381	9075
KING SALMON	313	322	513	908	1290	1606	1600	1333	1411	966	673	408	11343
KOTZEBUE	381	446	723	1249	1728	2127	2192	1932	2080	1554	1057	636	16105
MCGRATH	208	338	633	1184	1791	2232	2294	1817	1758	1122	648	258	14283
NOME	481	496	693	1094	1455	1820	1879	1666	1770	1314	930	573	14171
SAINT PAUL	605	539	612	862	963	1197	1228	1168	1265	1098	936	726	11199
SHEMYA	577	475	501	784	876	1042	1045	958	1011	885	837	696	9687
YAKUTAT	338	347	474	716	936	1144	1169	1019	1042	840	632	435	9092
ARIZ. FLAGSTAFF	46	68	201	558	867	1073	1169	991	911	651	437	180	7152
PHOENIX	0	0	0	22	234	415	474	328	217	75	0	0	1765
PRESCOTT	0	0	27	245	579	797	865	711	605	360	158	15	4362
TUCSON	0	0	0	25	231	406	471	344	242	75	6	0	1800
WINSLOW	0	0	6	245	711	1008	1054	770	601	291	96	0	4782
YUMA	0	0	0	0	148	319	363	228	130	29	0	0	1217
ARK. FORT SMITH	0	0	12	127	450	704	781	596	456	144	22	0	3292
LITTLE ROCK	0	0	9	127	465	716	756	577	434	126	9	0	3219
TEXARKANA	0	0	0	78	345	561	626	468	350	105	0	0	2533
CALIF. BAKERSFIELD	0	0	0	37	282	502	546	364	267	105	19	0	2122
BISHOP	0	0	42	248	576	797	874	666	539	306	143	36	4227
BLUE CANYON	34	50	120	347	579	766	865	781	791	582	397	195	5507
BURBANK	0	0	6	43	177	301	366	277	239	138	81	18	1646
EUREKA	270	257	258	329	414	499	546	470	505	438	372	285	4643
FRESNO	0	0	0	78	339	558	586	406	319	150	56	0	2492
LONG BEACH	0	0	12	40	156	288	375	297	267	168	90	18	1711
LOS ANGELES	28	22	42	78	180	291	372	302	288	219	158	81	2061
MT. SHASTA	25	34	123	406	696	902	983	784	738	525	347	159	5722
OAKLAND	53	50	45	127	309	481	527	400	353	255	180	90	2870
POINT ARGUELLO	202	186	162	205	291	400	474	392	403	339	298	243	3595
RED BLUFF	0	0	0	53	318	555	605	428	341	168	47	0	2515
SACRAMENTO	0	0	12	81	363	577	614	442	360	216	102	6	2773
SANDBERG	0	0	30	202	480	691	778	661	620	426	264	57	4209
SAN DIEGO	6	0	15	37	123	251	313	249	202	123	84	36	1439
SAN FRANCISCO	81	78	60	143	306	462	508	395	363	279	214	126	3015
SANTA CATALINA	16	0	9	50	165	279	353	308	326	249	192	105	2052
SANTA MARIA	99	93	96	146	270	391	459	370	363	282	233	165	2967
COLO. ALAMOSA	65	99	279	639	1065	1420	1476	1162	1020	696	440	168	8529
COLORADO SPRINGS	9	25	132	456	825	1032	1128	938	893	582	319	84	6423
DENVER	6	9	117	428	819	1035	1132	938	887	558	288	66	6283
GRAND JUNCTION	0	0	30	313	786	1113	1209	907	729	387	146	21	5641
PUEBLO	0	0	54	326	750	986	1085	871	772	429	174	15	5462
CONN. BRIDGEPORT	0	0	66	307	615	986	1079	966	853	510	208	27	5617
HARDFORT	0	6	99	372	711	1119	1209	1061	899	495	177	24	6172
NEW HAVEN	0	12	87	347	648	1011	1097	991	871	543	245	45	5897
DEL. WILMINGTON	0	0	51	270	588	927	980	874	735	387	112	6	4930
FLA. APALACHICOLA	0	0	0	16	153	319	347	260	180	33	0	0	1308
DAYTONA BEACH	0	0	0	0	75	211	248	190	140	15	0	0	879
FORT MYERS	0	0	0	0	24	109	146	101	62	0	0	0	442
JACKSONVILLE	0	0	0	12	144	310	332	246	174	21	0	0	1239
KEY WEST	0	0	0	0	0	28	40	31	9	0	0	0	108
LAKELAND	0	0	0	0	57	164	195	146	99	0	0	0	661
MIAMI BEACH	0	0	0	0	0	40	56	36	9	0	0	0	141
ORLANDO	0	0	0	0	72	198	220	165	105	6	0	0	766
PENSACOLA	0	0	0	19	195	353	400	277	183	36	0	0	1463
TALLAHASSEE	0	0	0	28	198	360	375	286	202	36	0	0	1485
TAMPA	0	0	0	0	60	171	202	148	102	0	0	0	683
WEST PALM BEACH	0	0	0	0	6	65	87	64	31	0	0	0	253

NORMAL TOTAL HEATING DEGREE DAYS (Base 65°)

STATE AND STATION	JULY	AUG.	SEP.	OCT.	NOV.	DEC.	JAN.	FEB.	MAR.	APR.	MAY	JUNE	ANNUAL
GA. ATHENS	0	0	12	115	405	632	642	529	431	141	22	0	2929
ATLANTA	0	0	18	127	414	626	639	529	437	168	25	0	2983
AUGUSTA	0	0	0	78	333	552	549	445	350	90	0	0	2397
COLUMBUS	0	0	0	87	333	543	552	434	338	96	0	0	2383
MACON	0	0	0	71	297	502	505	403	295	63	0	0	2136
ROME	0	0	24	161	474	701	710	577	468	177	34	0	3326
SAVANNAH	0	0	0	47	246	437	437	353	254	45	0	0	1819
THOMASVILLE	0	0	0	25	198	366	394	305	208	33	0	0	1529
IDAHO BOISE	0	0	132	415	792	1017	1113	854	722	438	245	81	5809
IDAHO FALLS 46W	16	34	270	623	1056	1370	1538	1249	1085	651	391	192	8475
IDAHO FALLS 42NW	16	40	282	648	1107	1432	1600	1291	1107	657	388	192	8760
LEWISTON	0	0	123	403	756	933	1063	815	694	426	239	90	5542
POCATELLO	0	0	172	493	900	1166	1324	1058	905	555	319	141	7033
ILL. CAIRO	0	0	36	164	513	791	856	680	539	195	47	0	3821
CHICAGO	0	0	81	326	753	1113	1209	1044	890	480	211	48	6155
MOLINE	0	9	99	335	774	1181	1314	1100	918	450	189	39	6408
PEORIA	0	6	87	326	759	1113	1218	1025	849	426	183	33	6025
ROCKFORD	6	9	114	400	837	1221	1333	1137	961	516	236	60	6830
SPRINGFIELD	0	0	72	291	696	1023	1135	935	769	354	136	18	5429
IND. EVANSVILLE	0	0	66	220	606	896	955	767	620	237	68	0	4435
FORT WAYNE	0	9	105	378	783	1135	1178	1028	890	471	189	39	6205
INDIANAPOLIS	0	0	90	316	723	1051	1113	949	809	432	177	39	5699
SOUTH BEND	0	6	111	372	777	1125	1221	1070	933	525	239	60	6439
IOWA Burlington	0	0	93	322	768	1135	1259	1042	859	426	177	33	6114
DES MOINES	0	9	99	363	837	1231	1398	1165	967	489	211	39	6808
DUBUQUE	12	31	156	450	906	1287	1420	1204	1026	546	260	78	7376
SIOUX CITY	0	9	108	369	867	1240	1435	1198	989	483	214	39	6951
WATERLOO	12	19	138	428	909	1296	1460	1221	1023	531	229	54	7320
KANS. CONCORDIA	0	0	57	276	705	1023	1163	935	781	372	149	18	5479
DODGE CITY	0	0	33	251	666	939	1051	840	719	354	124	9	4986
GOODLAND	0	6	81	381	810	1073	1166	955	884	507	236	42	6141
TOPEKA	0	0	57	270	672	980	1122	893	722	330	124	12	5182
WICHITA	0	0	33	229	618	905	1023	804	645	270	87	6	4620
KY. COVINGTON	0	0	75	291	669	983	1035	893	756	390	149	24	5265
LEXINGTON	0	0	54	239	609	902	946	818	685	325	105	0	4683
LOUISVILLE	0	0	54	248	609	890	930	818	682	315	105	9	4660
LA. ALEXANDRIA	0	0	0	56	273	431	471	361	260	69	0	0	1921
BATON ROUGE	0	0	0	31	216	369	409	294	208	33	0	0	1560
BURRWOOD	0	0	0	0	96	214	298	218	171	27	0	0	1024
LAKE CHARLES	0	0	0	19	210	341	381	274	195	39	0	0	1459
NEW ORLEANS	0	0	0	19	192	322	363	258	192	39	0	0	1385
SHREVEPORT	0	0	0	47	297	477	552	426	304	81	0	0	2184
MAINE CARIBOU	78	115	336	682	1044	1535	1690	1470	1308	858	468	183	9767
PORTLAND	12	53	195	508	807	1215	1339	1182	1042	675	372	111	7511
MD. BALTIMORE	0	0	48	264	585	905	936	820	679	327	90	0	4654
FREDERICK	0	0	66	307	624	955	995	876	741	384	127	12	5087
MASS. BLUE HILL OBSY	0	22	108	381	690	1085	1178	1053	936	579	267	69	6368
BOSTON	0	9	60	316	603	983	1088	972	846	513	208	36	5634
NANTUCKET	12	22	93	332	573	896	992	941	896	621	384	129	5891
PITTSFIELD	25	59	219	524	831	1231	1339	1196	1063	660	326	105	7578
WORCESTER	6	34	147	450	774	1172	1271	1123	998	612	304	78	6969
MICH. ALPENA	68	105	273	580	912	1268	1404	1299	1218	777	446	156	8506
DETROIT (CITY)	0	0	87	360	738	1088	1181	1058	936	522	220	42	6232
ESCANABA	59	87	243	539	924	1293	1445	1296	1203	777	456	159	8481
FLINT	16	40	159	465	843	1212	1330	1198	1066	639	319	90	7377
GRAND RAPIDS	9	28	135	434	804	1147	1259	1134	1011	579	279	75	6894
LANSING	6	22	138	431	813	1163	1262	1142	1011	579	273	69	6909
MARQUETTE	59	81	240	527	936	1268	1411	1268	1187	771	468	177	8393
MUSKEGON	12	28	120	400	762	1088	1209	1100	995	594	310	78	6696
SAULT STE. MARIE	96	105	279	580	951	1367	1525	1380	1277	810	477	201	9048
MINN. DULUTH	71	109	330	632	1131	1581	1745	1518	1355	840	490	198	10000
INTERNATIONAL FALLS	71	112	363	701	1236	1724	1919	1621	1414	828	443	174	10606
MINNEAPOLIS	22	31	189	505	1014	1454	1631	1380	1166	621	288	81	8382
ROCHESTER	25	34	186	474	1005	1438	1593	1366	1150	630	301	93	8295
SAINT CLOUD	28	47	225	549	1065	1500	1702	1445	1221	666	326	105	8879

NORMAL TOTAL HEATING DEGREE DAYS (Base 65°)

STATE AND STATION	JULY	AUG.	SEP	OCT.	NOV.	DEC.	JAN.	FEB.	MAR.	APR.	MAY	JUNE	ANNUAL
MISS. JACKSON	0	0	0	65	315	502	546	414	310	87	0	0	2239
MERIDIAN	0	0	0	81	339	518	543	417	310	81	0	0	2289
VICKSBURG	0	0	0	53	279	462	512	384	282	69	0	0	2041
MO. COLUMBIA	0	0	54	251	651	967	1076	874	716	324	121	12	5046
KANSAS	0	0	39	220	612	905	1032	818	682	294	109	0	4711
ST. JOSEPH	0	6	60	285	708	1039	1172	949	769	348	133	15	5484
ST. LOUIS	0	0	60	251	627	936	1026	848	704	312	121	15	4900
SPRINGFIELD	0	0	45	223	600	877	973	781	660	291	105	6	4561
MONT. BILLINGS	6	15	186	487	897	1135	1296	1100	970	570	285	102	7049
GLASGOW	31	47	270	608	1104	1466	1711	1439	1187	648	335	150	8996
GREAT FALLS	28	53	258	543	921	1169	1349	1154	1063	642	384	186	7750
HAVRE	28	53	306	595	1065	1367	1584	1364	1181	657	338	162	8700
HELENA	31	59	294	601	1002	1265	1438	1170	1042	651	381	195	8129
KALISPELL	50	99	321	654	1020	1240	1401	1134	1029	639	397	207	8191
MILES CITY	6	6	174	502	972	1296	1504	1252	1057	579	276	99	7723
MISSOULA	34	74	303	651	1035	1287	1420	1120	970	621	391	219	8125
NEBR. GRAND ISLAND	0	6	108	381	834	1172	1314	1089	908	462	211	45	6530
LINCOLN	0	6	75	301	726	1066	1237	1016	834	402	171	30	5864
NORFOLK	9	0	111	397	873	1234	1414	1179	983	498	233	48	6979
NORTH PLATTE	0	6	123	440	885	1166	1271	1039	930	519	248	57	6684
OMAHA	0	12	105	357	828	1175	1355	1126	939	465	208	42	6612
SCOTTSBLUFF	0	0	138	459	876	1128	1231	1008	921	552	285	75	6673
VALENTINE	9	12	165	493	942	1237	1395	1176	1045	579	288	84	7425
NEV. ELKO	9	34	225	561	924	1197	1314	1036	911	621	409	192	7433
ELY	28	43	234	592	939	1184	1308	1075	977	672	456	225	7733
LAS VEGAS	0	0	0	78	387	617	688	487	335	111	6	0	2709
RENO	43	87	204	490	801	1026	1073	823	729	510	357	189	6332
WINNEMUCCA	0	34	210	536	876	1091	1172	916	837	573	363	153	6761
N. H. CONCORD	6	50	177	505	822	1240	1358	1184	1032	636	298	75	7383
MT. WASH. OBSY.	493	536	720	1057	1341	1742	1820	1663	1652	1260	930	603	13817
N. J. ATLANTIC CITY	0	0	39	251	549	880	936	848	741	420	133	15	4812
NEWARK	0	0	30	248	573	921	983	876	729	381	118	0	4859
TRENTON	0	0	57	264	576	924	989	885	753	399	121	12	4980
N. MEX. ALBUQUERQUE	0	0	12	229	642	868	930	703	595	288	81	0	4348
CLAYTON	0	6	66	310	699	899	986	812	747	429	183	21	5158
RATON	9	28	126	431	825	1048	1116	904	834	543	301	63	6228
ROSWELL	0	0	18	202	573	806	840	641	481	201	31	0	3793
SILVER CITY	0	0	6	183	525	729	791	605	518	261	87	0	3705
N. Y. ALBANY	0	19	138	440	777	1194	1311	1156	992	564	239	45	6875
BINGHAMTON (AP)	22	65	201	471	810	1184	1277	1154	1045	645	313	99	7286
BINGHAMTON (PO)	0	28	141	406	732	1107	1190	1081	949	543	229	45	6451
BUFFALO	19	37	141	440	777	1156	1256	1145	1039	645	329	78	7062
CENTRAL PARK	0	0	30	233	540	902	986	885	760	408	118	9	4871
J. F. KENNEDY INTL.	0	0	36	248	564	933	1029	935	815	480	167	12	5219
LAGUARDIA	0	0	27	223	528	887	973	879	750	414	124	6	4811
ROCHESTER	9	31	126	415	747	1125	1234	1123	1014	597	279	48	6748
SCHENECTADY	0	22	123	422	756	1159	1283	1131	970	543	211	30	6650
SYRACUSE	6	28	132	415	744	1153	1271	1140	1004	570	248	45	6756
N.C. ASHEVILLE	0	0	48	245	555	775	784	683	592	273	87	0	4042
CAPE HATTERAS	0	0	0	78	273	521	580	518	440	177	25	0	2612
CHARLOTTE	0	0	6	124	438	691	691	582	481	156	22	0	3191
GREENSBORO	0	0	33	192	513	778	784	672	552	234	47	0	3805
RALEIGH	0	0	21	164	450	716	725	616	487	180	34	0	3393
WILMINGTON	0	0	0	74	291	521	546	462	357	96	0	0	2347
WINSTON SALEM	0	0	21	171	483	747	753	652	524	207	37	0	3595
N. DAK. BISMARCK	34	28	222	577	1083	1463	1708	1442	1203	645	329	117	8851
DEVILS LAKE	40	53	273	642	1191	1634	1872	1579	1345	753	381	138	9901
FARGO	28	37	219	574	1107	1569	1789	1520	1262	690	332	99	9226
WILLISTON	31	43	261	601	1122	1513	1758	1473	1262	681	357	141	9243
OHIO AKRON	0	9	96	381	726	1070	1138	1016	871	489	202	39	6037
CINCINNATI	0	0	54	248	612	921	970	837	701	336	118	9	4806
CLEVELAND	9	25	105	384	738	1088	1159	1047	918	552	260	66	6351
COLUMBUS	0	6	84	347	714	1039	1088	949	809	426	171	27	5660
DAYTON	0	6	78	310	696	1045	1097	955	809	429	167	30	5622
MANSFIELD	9	22	114	397	768	1110	1169	1042	924	543	245	60	6403
SANDUSKY	0	6	66	313	684	1032	1107	991	868	495	198	36	5796
TOLEDO	0	16	117	406	792	1138	1200	1056	924	543	242	60	6494
YOUNGSTOWN	6	19	120	412	771	1104	1169	1047	921	540	248	60	6417

NORMAL TOTAL HEATING DEGREE DAYS (Base 65°)

STATE AND STATION	JULY	AUG.	SEP.	OCT.	NOV.	DEC.	JAN.	FEB.	MAR.	APR.	MAY	JUNE	ANNUAL
OKLA. OKLAHOMA CITY	0	0	15	164	498	766	868	664	527	189	34	0	3725
TULSA	0	0	18	158	522	787	893	683	539	213	47	0	3860
OREG. ASTORIA	146	130	210	375	561	679	753	622	636	480	363	231	5186
BURNS	12	37	210	515	867	1113	1246	988	856	570	366	177	6957
EUGENE	34	34	129	366	585	719	803	627	589	426	279	135	4726
MEACHAM	84	124	288	580	918	1091	1209	1005	983	726	527	339	7874
MEDFORD	0	0	78	372	678	871	918	697	642	432	242	78	5008
PENDLETON	0	0	111	350	711	884	1017	773	617	396	205	63	5127
PORTLAND	25	28	114	335	597	735	825	644	586	396	245	105	4635
ROSEBURG	22	16	105	329	567	713	766	608	570	405	267	123	4491
SALEM	37	31	111	338	594	729	822	647	611	417	273	144	4754
SEXTON SUMMIT	81	81	171	443	666	874	958	809	818	609	465	279	6254
PA. ALLENTOWN	0	0	90	353	693	1045	1116	1002	849	471	167	24	5810
ERIE	0	25	102	391	714	1063	1169	1081	973	585	288	60	6451
HARRISBURG	0	0	63	298	648	992	1045	907	766	396	124	12	5251
PHILADELPHIA	0	0	60	291	621	964	1014	890	744	390	115	12	5101
PITTSBURGH	0	9	105	375	726	1063	1119	1002	874	480	195	39	5987
READING	0	0	54	257	597	939	1001	885	735	372	105	0	4945
SCRANTON	0	19	132	434	762	1104	1156	1028	893	498	195	33	6254
WILLIAMSPORT	0	9	111	375	717	1073	1122	1002	856	468	177	24	5934
R. I. BLOCK IS.	0	16	78	307	594	902	1020	955	877	612	344	99	5804
PROVIDENCE	0	16	96	372	660	1023	1110	988	868	534	236	51	5954
S. C. CHARLESTON	0	0	0	59	282	471	487	389	291	54	0	0	2033
COLUMBIA	0	0	0	84	345	577	570	470	357	81	0	0	2484
FLORENCE	0	0	0	78	315	552	552	459	347	84	0	0	2387
GREENVILLE	0	0	0	112	387	636	648	535	434	120	12	0	2884
SPARTANBURG	0	0	15	130	417	667	663	560	453	144	25	0	3074
S. DAK. HURON	9	12	165	508	1014	1432	1628	1355	1125	600	288	87	8223
RAPID CITY	22	12	165	481	897	1172	1333	1145	1051	615	326	126	7345
SIOUX FALLS	19	25	168	462	972	1361	1544	1285	1082	573	270	78	7839
TENN. BRISTOL	0	0	51	236	573	828	828	700	598	261	68	0	4143
CHATTANOOGA	0	0	18	143	468	698	722	577	453	150	25	0	3254
KNOXVILLE	0	0	30	171	489	725	732	613	493	198	43	0	3494
MEMPHIS	0	0	18	130	447	698	729	585	456	147	22	0	3232
NASHVILLE	0	0	30	158	495	732	778	644	512	189	40	0	3578
OAK RIDGE (CO)	0	0	39	192	531	772	778	669	552	228	56	0	3817
TEX. ABILENE	0	0	0	99	366	586	642	470	347	114	0	0	2624
AMARILLO	0	0	18	205	570	797	877	664	546	252	56	0	3985
AUSTIN	0	0	0	31	225	388	468	325	223	51	0	0	1711
BROWNSVILLE	0	0	0	0	66	149	205	106	74	0	0	0	600
CORPUS CHRISTI	0	0	0	0	120	220	291	174	109	0	0	0	914
DALLAS	0	0	0	62	321	524	601	440	319	90	6	0	2363
EL PASO	0	0	0	84	414	648	685	445	319	105	0	0	2700
FORT WORTH	0	0	0	65	324	536	614	448	319	99	0	0	2405
GALVESTON	0	0	0	0	138	270	350	258	189	30	0	0	1235
HOUSTON	0	0	0	6	183	307	384	288	192	36	0	0	1396
LAREDO	0	0	0	0	105	217	267	134	74	0	0	0	797
LUBBOCK	0	0	18	174	513	744	800	613	484	201	31	0	3578
MIDLAND	0	0	0	87	381	592	651	468	322	90	0	0	2591
PORT ARTHUR	0	0	0	22	207	329	384	274	192	39	0	0	1447
SAN ANGELO	0	0	0	68	318	536	567	412	288	66	0	0	2255
SAN ANTONIO	0	0	0	31	207	363	428	286	195	39	0	0	1549
VICTORIA	0	0	0	6	150	270	344	230	152	21	0	0	1173
WACO	0	0	0	43	270	456	536	389	270	66	0	0	2030
WICHITA FALLS	0	0	0	99	381	632	698	518	378	120	6	0	2832
UTAH MILFORD	0	0	99	443	867	1141	1252	988	822	519	279	87	6497
SALT LAKE CITY	0	0	81	419	849	1082	1172	910	763	459	233	84	6052
WENDOVER	0	0	48	372	822	1091	1178	902	729	408	177	51	5778
VT. BURLINGTON	28	65	207	539	891	1349	1513	1333	1187	714	353	90	8269
VA. CAPE HENRY	0	0	0	112	360	645	694	633	536	246	53	0	3279
LYNCHBURG	0	0	51	223	540	822	849	731	605	267	78	0	4166
NORFOLK	0	0	0	136	408	698	738	655	533	216	37	0	3421
RICHMOND	0	0	36	214	495	784	815	703	546	219	53	0	3865
ROANOKE	0	0	51	229	549	825	834	722	614	261	65	0	4150
WASH. NAT'L. AP.	0	0	33	217	519	834	871	762	626	288	74	0	4224

NORMAL TOTAL HEATING DEGREE DAYS (Base 65°)

STATE AND STATION	JULY	AUG.	SEP.	OCT.	NOV.	DEC.	JAN.	FEB.	MAR.	APR.	MAY	JUNE	ANNUAL
WASH. OLYMPIA	68	71	198	422	636	753	834	675	645	450	307	177	5236
SEATTLE	50	47	129	329	543	657	738	599	577	396	242	117	4424
SEATTLE BOEING	34	40	147	384	624	763	831	655	608	411	242	99	4838
SEATTLE TACOMA	56	62	162	391	633	750	828	678	657	474	295	159	5145
SPOKANE	9	25	168	493	879	1082	1231	980	834	531	288	135	6655
STAMPEDE PASS	273	291	393	701	1008	1178	1287	1075	1085	855	654	483	9283
TATOOSH IS.	295	279	306	406	534	639	713	613	645	525	431	333	5719
WALLA WALLA	0	0	87	310	681	843	986	745	589	342	177	45	4805
YAKIMA	0	12	144	450	828	1039	1163	868	713	435	220	69	5941
W. VA. CHARLESTON	0	0	63	254	591	865	880	770	648	300	96	9	4476
ELKINS	9	25	135	400	729	992	1008	896	791	444	198	48	5675
HUNTINGTON	0	0	63	257	585	856	880	764	636	294	99	12	4446
PARKERSBURG	0	0	60	264	606	905	942	826	691	339	115	6	4754
WIS. GREEN BAY	28	50	174	484	924	1333	1494	1313	1141	654	335	99	8029
LA CROSSE	12	19	153	437	924	1339	1504	1277	1070	540	245	69	7589
MADISON	25	40	174	474	930	1330	1473	1274	1113	618	310	102	7863
MILWAUKEE	43	47	174	471	876	1252	1376	1193	1054	642	372	135	7635
WYO. CASPER	6	16	192	524	942	1169	1290	1084	1020	657	381	129	7410
CHEYENNE	19	31	210	543	924	1101	1228	1056	1011	672	381	102	7278
LANDER	6	19	204	555	1020	1299	1417	1145	1017	654	381	153	7870
SHERIAN	25	31	219	539	948	1200	1355	1154	1054	642	366	150	7683

One of the most practical of weather statistics is the "heating degree day." First devised some 50 years ago, the degree day system has been in quite general use by the heating industry for more than 30 years.

Heating degree days are the number of degrees the daily average temperature is below 65°. Normally heating is not required in a building when the outdoor average daily temperature is 65°. Heating degree days are determined by substracting the average daily temperatures below 65° from the base 65°. A day with an average temperature of 50° has 15 heating degree days (65 — 50 = 15) while one with an average temperature of 65° or higher has none.

Several characteristics make the degree day figures especially useful. They are cumulative so that the degree day sum for a period of days represents the total heating load for that period. The relationship between degree days and fuel consumption is linear, i.e., doubling the degree days usually doubles the fuel consumption. Comparing normal seasonal degree days in different locations gives a rough estimate of seasonal fuel consumption. For example, it would require roughly 4½ times as much fuel to heat a building in Chicago, Ill., where the mean annual total heating degree days are about 6,200 than to heat a similar building in New Orleans, La., where the annual total heating degree days are around 1,400. Using degree days has the advantage that the consumption ratios are fairly constant, i.e., the fuel consumed per 100 degree days is about the same whether the 100 degree days occur in only 3 or 4 days or are spread over 7 or 8 days.

The rapid adoption of the degree day system paralleled the spread of automatic fuel systems in the 1930's. Since oil and gas are more costly to store than solid fuels, this places a premium on the scheduling of deliveries and the precise evaluation of use rates and peak demands.

PRECIPITATION OF 0.01 INCH OR MORE

STATE AND STATION	YEARS	JAN.	FEB.	MAR.	APR.	MAY	JUNE	JULY	AUG.	SEPT.	OCT.	NOV.	DEC.	ANNUAL
ALA. BIRMINGHAM	18	11	11	12	9	9	10	12	10	8	6	9	11	118
MOBILE	20	10	11	11	8	8	12	18	13	10	6	8	11	125
MONTGOMERY U	83	11	10	10	9	9	11	12	11	8	6	7	10	114
ALASKA ANCHORAGE	35	7	6	6	5	5	7	11	15	14	10	8	7	104
BARROW	41	4	4	3	3	3	4	8	10	9	10	6	5	69
BETHEL	18	11	11	13	9	11	12	16	21	16	13	13	13	158
COLD BAY	15	17	15	17	12	16	15	17	19	19	22	20	19	207
FAIRBANKS	32	8	7	7	4	7	10	13	14	10	11	9	7	108
FT. YUKON	35	6	5	5	2	3	6	7	9	8	7	5	5	68
JUNEAU	18	18	17	18	17	17	15	17	18	20	23	20	22	222
NOME	39	10	9	10	9	8	9	14	18	15	11	10	10	133
ST. PAUL ISLAND	43	17	14	14	13	13	12	15	19	20	22	21	19	200
ARIZ. FLAGSTAFF	12	6	6	6	5	3	3	12	11	4	4	5	5	70
PHOENIX	22	4	4	3	2	1	1	4	5	3	3	2	4	34
TUCSON	20	4	4	4	2	1	1	11	9	3	4	2	4	49
WINSLOW	30	4	4	4	3	3	2	7	9	5	4	3	4	53
YUMA	11	3	2	2	1	*	*	1	2	*	1	1	2	14
ARK. FT. SMITH	16	8	9	9	10	11	8	8	7	7	6	6	7	96
LITTLE ROCK	19	10	10	11	11	10	8	8	7	7	6	8	9	105
CALIF. BAKERSFIELD	24	6	7	7	4	2	*	*	*	1	2	3	6	37
EUREKA	51	17	14	15	12	9	5	2	2	5	9	12	15	118
FRESNO	21	8	7	7	5	2	1	*	*	1	2	4	8	46
LOS ANGELES U	21	6	5	6	4	2	1	*	1	1	2	4	5	36
RED BLUFF	17	12	9	9	6	6	3	1	1	2	5	7	10	71
SACRAMENTO	28	11	10	8	6	3	1	*	*	1	3	6	9	58
SAN DIEGO	21	7	6	7	5	2	1	*	*	1	2	4	6	42
SAN FRANCISCO	34	11	10	9	6	3	1	*	*	1	4	6	10	62
COLO. ALAMOSA	16	4	4	4	5	7	4	9	10	5	5	4	3	64
DENVER	27	6	6	8	9	11	9	9	8	6	6	5	4	87
GRAND JUNCTION	15	8	7	8	6	6	4	4	7	5	5	5	6	71
PUEBLO	21	5	5	6	7	9	7	9	9	4	4	3	3	72
CONN. HARTFORD U	51	12	11	12	12	12	11	10	10	9	8	10	10	127
NEW HAVEN	18	12	11	13	13	13	11	11	10	9	8	11	11	132
DEL. WILMINGTON	14	12	10	13	12	13	10	9	9	8	9	9	10	123
D. C. WASHINGTON	20	11	8	12	10	12	9	10	10	8	7	8	9	115
FLA. APALACHICOLA	31	8	8	8	6	6	10	16	14	12	6	6	8	108
JACKSONVILLE	20	7	8	8	7	8	11	15	14	14	8	6	8	115
KEY WEST	13	6	5	6	5	7	11	14	15	15	13	7	8	113
MIAMI	19	6	5	6	7	10	13	16	16	18	15	8	7	127
ORLANDO	19	5	6	7	7	8	14	18	16	15	9	5	6	117
TAMPA	15	6	7	8	7	6	12	18	17	15	8	5	6	113
GA. ATLANTA U	77	12	11	12	10	9	11	12	11	8	7	8	11	122
AUGUSTA U	85	10	10	10	8	9	11	12	12	7	6	7	10	112
MACON	13	9	10	11	8	9	10	13	11	9	6	6	9	111
SAVANNAH U	85	9	9	9	7	9	12	14	13	10	7	6	9	114
THOMASVILLE	39	9	9	9	7	9	13	16	14	10	5	6	9	116
HAWAII HILO	19	19	19	23	25	25	24	28	27	24	25	24	24	287
HONOLULU	39	12	12	13	12	11	11	13	13	12	12	13	14	148
LIHUE	10	15	16	17	15	16	17	20	20	16	19	17	18	204
IDAHO BOISE	22	12	11	10	8	9	6	2	2	3	7	10	11	90
POCATELLO	23	11	11	10	7	9	6	4	4	5	6	8	11	92
ILL. CAIRO U	19	11	10	12	12	12	10	9	8	8	7	9	9	117
CHICAGO U	85	11	10	12	11	12	11	9	9	9	9	10	11	124
SPRINGFIELD	14	9	9	12	12	11	11	10	8	7	7	9	9	114
IND. EVANSVILLE	21	11	9	12	12	11	10	10	8	7	7	9	10	116
FT. WAYNE	15	13	11	13	14	12	11	10	9	8	8	10	12	131
INDIANAPOLIS	22	12	10	12	12	13	11	9	8	7	8	10	11	123
IOWA BURLINGTON U	64	7	7	10	10	12	11	8	8	9	7	7	7	103
DES MOINES	22	7	8	10	10	11	11	9	9	8	7	6	7	103
DUBUQUE	11	9	8	11	11	11	10	10	9	8	8	7	9	111
SIOUX CITY	20	6	7	9	9	11	12	9	9	7	6	5	7	97
KANS. CONCORDIA U	76	5	6	7	9	11	11	8	7	7	6	4	5	86
DODGE CITY	19	4	5	7	7	10	9	9	9	6	5	4	3	78
GOODLAND	41	4	5	6	8	10	10	8	8	5	5	4	4	77
WICHITA U	67	5	6	7	8	11	9	8	8	8	7	5	5	87
KY. LEXINGTON	17	14	11	13	13	12	11	11	9	7	7	11	11	130
LOUISVILLE	14	13	11	13	13	11	11	10	8	7	7	9	11	123

PRECIPITATION OF 0.01 INCH OR MORE

STATE AND STATION	YEARS	JAN.	FEB.	MAR.	APR.	MAY	JUNE	JULY	AUG.	SEPT.	OCT.	NOV.	DEC.	ANNUAL
LA. LAKE CHARLES	23	10	10	8	8	8	9	12	10	8	5	8	10	106
NEW ORLEANS	46	10	9	9	7	9	12	15	14	10	7	7	10	119
SHREVEPORT U	84	11	9	9	9	9	8	8	7	6	6	8	9	99
MAINE CARIBOU	22	14	14	13	14	13	15	14	13	12	12	14	14	162
PORTLAND	21	12	11	11	12	13	11	9	9	8	9	11	10	126
MD. BALTIMORE U	91	11	10	12	11	11	11	11	11	8	8	9	10	123
MASS. BOSTON U	85	12	11	12	11	11	10	10	10	9	9	10	11	126
NANTUCKET	15	13	11	13	12	11	8	8	9	7	9	12	12	125
MICH. ALPENA U	45	15	12	12	11	11	11	9	10	12	12	14	15	144
DETROIT	28	13	12	13	13	12	11	9	9	9	9	11	13	134
ESCANABA U	52	11	10	10	10	11	11	11	11	12	10	11	11	129
GRAND RAPIDS U	52	14	12	12	12	12	10	8	9	10	10	12	14	135
MARQUETTE	24	17	15	12	11	12	12	10	11	13	11	16	15	155
S. STE. MARIE	20	18	15	13	11	11	12	10	10	13	11	18	19	161
MINN. DULUTH U	85	10	9	10	9	12	13	11	11	11	10	10	10	126
INTERNATIONAL FALLS	22	12	10	10	10	12	13	11	12	12	9	12	12	135
MINNEAPOLIS	23	8	7	11	9	11	12	10	10	9	8	8	9	112
MISS. MERIDIAN	16	10	10	10	8	9	8	11	9	7	5	7	9	105
VICKSBURG	24	10	10	11	9	8	9	10	8	7	5	8	10	105
MO. KANSAS CITY	28	7	7	9	11	12	11	8	8	8	7	6	6	100
ST. LOUIS U	23	9	9	11	12	12	11	8	8	7	7	8	8	110
SPRINGFIELD	16	8	9	10	10	12	10	10	8	7	7	8	8	107
MONT. BILLINGS	27	7	7	9	9	10	11	7	6	7	6	6	6	92
HAVRE U	57	8	7	7	7	10	12	8	7	7	5	6	6	90
HELENA	21	7	7	8	8	11	12	8	8	6	6	7	8	97
KALISPELL	12	16	12	12	9	11	12	6	8	8	11	11	15	132
MILES CITY	24	7	6	7	7	10	11	8	7	7	5	5	6	88
NEBR. NORTH PLATTE	10	4	6	7	8	11	9	8	8	5	5	5	4	80
OMAHA	26	6	7	8	9	11	11	9	10	8	6	5	6	96
SCOTTS BLUFF	18	4	5	8	8	13	11	8	7	6	5	4	4	83
VALENTINE U	67	6	6	8	9	11	12	9	9	7	5	5	6	93
NEV. LAS VEGAS	19	3	3	3	2	1	*	3	2	2	2	2	3	26
RENO	19	6	6	6	4	5	3	3	2	2	3	4	5	47
WINNEMUCCA U	77	9	9	8	7	7	5	2	2	3	5	6	9	72
N. H. CONCORD U	53	11	9	10	11	10	10	10	10	10	9	10	10	120
MT. WASHINGTON R	29	18	17	19	18	18	16	17	15	15	14	19	19	205
N. J. ATLANTIC CITY U	85	12	10	12	11	11	10	10	10	8	9	9	10	122
N. MEX. ALBUQUERQUE	22	4	4	4	4	4	3	9	10	5	5	3	4	58
ROSWELL U	51	3	3	3	4	5	5	8	7	6	4	3	3	54
N. Y. ALBANY U	82	12	11	12	12	12	12	12	11	10	10	11	11	136
BINGHAMTON U	65	15	13	14	14	13	12	12	11	10	11	13	14	152
BUFFALO U	85	19	17	16	13	13	11	10	9	11	12	15	18	164
NEW YORK U	90	12	10	12	11	11	10	11	10	9	9	9	10	124
SYRACUSE	12	19	16	18	17	13	11	12	10	10	11	16	19	171
N. C. ASHEVILLE.	31	11	11	12	10	11	12	15	12	8	8	8	10	129
CAPE HATTERAS R	81	12	10	11	9	9	10	12	12	9	8	9	11	122
CHARLOTTE	22	9	10	12	9	9	9	12	10	7	7	7	10	111
RALEIGH	17	10	11	10	9	10	10	12	11	9	8	9	9	117
WILMINGTON U	85	9	10	10	8	9	12	14	14	10	7	7	10	120
N. DAK. BISMARCK	22	8	7	8	7	9	12	9	9	7	6	6	7	95
FARGO	20	8	7	8	8	10	11	10	9	9	6	7	8	101
WILLISTON U	45	8	6	7	7	8	12	9	7	7	5	6	7	89
OHIO AKRON-CANTON	13	17	14	16	16	12	11	11	9	8	9	13	14	152
CINCINNATI (ABBE)	46	13	11	13	13	13	13	10	9	9	9	10	11	134
CLEVELAND	20	16	14	16	15	14	11	10	9	9	10	15	15	154
COLUMBUS	22	14	12	14	14	13	12	11	9	8	8	11	12	137
OKLA. OKLAHOMA CITY	21	6	7	7	8	11	9	7	6	6	6	4	5	82
TULSA	23	6	8	8	9	11	9	7	7	6	7	6	7	91
OREG. ASTORIA	8	24	21	22	18	16	15	7	9	11	18	18	22	201
BURNS	19	12	10	9	7	8	7	3	3	3	7	10	11	90
PORTLAND U	59	19	16	17	14	12	9	3	4	7	12	17	19	152
ROSEBURG U	78	18	16	16	13	11	7	2	2	6	11	15	17	134
PA. HARRISBURG	23	12	9	12	12	13	11	10	10	8	8	9	10	125
PHILADELPHIA	21	11	9	12	11	12	10	10	10	8	8	9	10	119
PITTSBURGH U	84	16	14	15	13	13	12	12	10	9	10	12	14	150
WILLIAMSPORT	17	13	12	14	14	14	11	12	11	9	10	13	13	144

PRECIPITATION OF 0.01 INCH OR MORE

STATE AND STATION	YEARS	JAN.	FEB.	MAR.	APR.	MAY	JUNE	JULY	AUG.	SEPT.	OCT.	NOV.	DEC.	ANNUAL
R. I. BLOCK ISLAND U	77	12	11	12	11	11	9	9	9	8	9	10	11	122
PROVIDENCE U	51	12	10	12	11	11	10	10	9	8	8	10	10	121
S. C. CHARLESTON U	85	9	9	9	7	8	11	13	12	10	6	7	9	110
COLUMBIA U	68	9	10	10	8	9	11	13	12	8	6	7	10	113
GREENVILLE U & AP	38	11	13	11	10	10	10	13	11	8	7	8	10	122
S. DAK. HURON U	74	7	6	8	9	10	11	9	9	7	6	5	7	94
RAPID CITY U	56	6	6	8	9	12	13	10	8	7	6	4	5	94
TENN. CHATTANOOGA U	77	13	12	13	11	11	12	13	11	8	7	9	12	132
KNOXVILLE U	84	13	11	13	11	12	12	12	11	8	7	9	12	131
MEMPHIS U	84	11	10	11	10	10	9	9	8	7	7	9	11	112
NASHVILLE	20	12	11	13	11	11	10	10	9	7	7	9	11	120
TEX. ABILENE	21	6	6	4	7	8	7	5	5	5	6	4	4	67
AMARILLO	20	4	4	4	6	9	7	9	8	5	5	3	4	68
BROWNSVILLE	19	7	7	5	4	4	5	4	7	10	6	7	6	72
CORPUS CHRISTI	22	8	8	6	5	6	5	4	5	8	7	6	7	75
DALLAS	21	8	8	8	9	8	6	5	6	5	6	6	6	81
EL PASO	22	3	2	2	2	2	4	8	7	4	4	2	3	43
HOUSTON	24	10	10	9	8	8	8	10	9	9	7	8	10	106
LAREDO	18	6	6	4	5	6	4	3	5	6	5	5	6	61
LUBBOCK	15	4	4	4	4	8	7	8	5	4	5	3	3	59
MIDLAND	26	4	3	2	3	6	4	5	4	4	6	3	4	48
SAN ANTONIO U	71	8	8	7	8	8	6	6	5	7	6	6	8	83
UTAH MILFORD	30	6	6	7	5	5	3	4	5	3	4	5	5	58
SALT LAKE CITY	33	10	9	10	9	8	5	4	6	5	6	7	9	86
VT. BURLINGTON U	49	13	12	13	13	13	12	12	11	11	11	13	13	147
VA. NORFOLK U	85	11	11	12	10	11	11	12	12	8	8	8	10	124
RICHMOND U	58	10	11	12	10	12	11	11	11	10	7	8	10	122
ROANOKE	14	11	11	12	10	12	11	11	12	9	8	9	9	124
WASH. SEATTLE U	64	18	16	16	13	12	9	5	5	8	13	18	19	152
SPOKANE U	74	14	12	11	9	9	8	4	4	6	8	12	15	112
STAMPEDE PASS	18	22	21	22	19	16	15	8	11	12	17	19	23	205
TATOOSH ISLAND R	59	22	18	20	17	14	12	10	10	11	17	21	23	197
WALLA WALLA U	47	13	11	12	9	9	7	3	3	5	9	11	14	106
YAKIMA U	46	9	7	5	4	5	5	2	2	3	5	8	9	64
W. VA. CHARLESTON	14	17	14	16	14	14	12	13	10	8	10	12	14	152
HUNTINGTON U	15	15	12	13	13	13	11	11	9	8	8	10	12	134
PARKERSBURG U	53	15	13	14	13	12	13	12	10	9	9	11	13	143
WIS. GREEN BAY U	69	10	9	10	11	12	11	10	9	10	9	9	10	120
LA CROSSE U	83	10	8	10	10	12	12	10	9	10	9	8	9	117
MADISON U	77	9	8	10	11	12	11	9	9	10	8	8	10	115
MILWAUKEE U	85	11	10	11	11	12	11	9	9	9	9	9	11	122
WYO. CASPER	22	7	7	10	10	11	8	7	5	6	6	7	6	92
CHEYENNE U	85	6	6	9	10	12	10	11	10	6	6	5	5	96
LANDER U	64	4	5	7	8	9	6	6	5	5	6	4	4	69
SHERIDAN	21	7	9	12	11	12	11	7	7	8	7	8	8	107
YELLOWSTONE	44	13	11	12	10	13	12	10	9	8	9	10	12	129
P. R. SAN JUAN	62	20	14	14	14	16	17	19	20	18	18	19	20	209

* LESS THAN ONCE IN 2 YEARS.

DATA FROM AIRPORT, EXCEPT THOSE MARKED WITH U FOR URBAN AND R FOR RURAL.

Charts and tabulation based on data generally for periods of record through 1961, from State and Local **Climatological Data.** Pattern too complex in Hawaii to indicate on small scale maps.

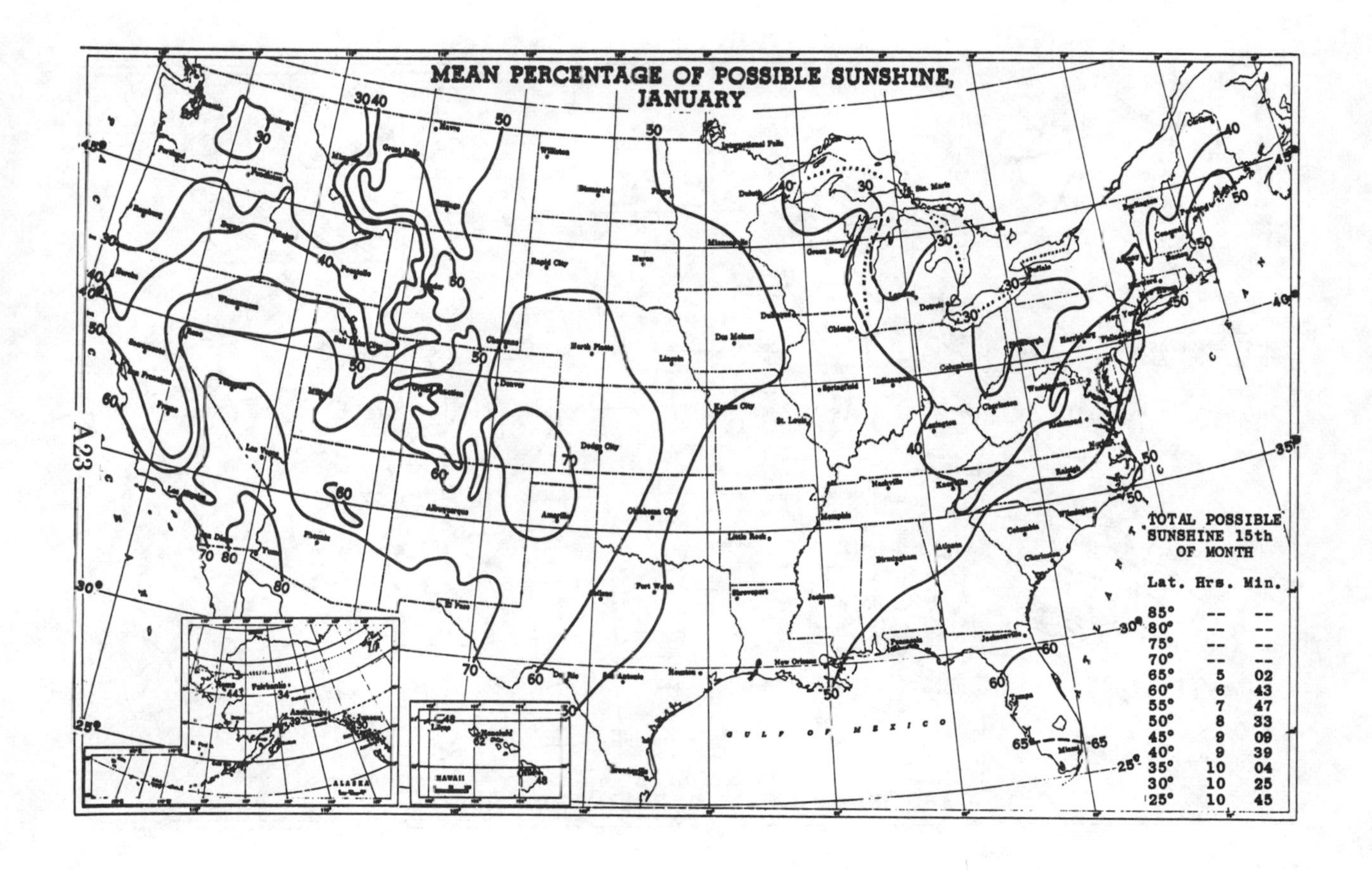

MEAN PERCENTAGE OF POSSIBLE SUNSHINE, JANUARY
TOTAL POSSIBLE SUNSHINE 15th OF MONTH
Lat. Hrs. Min.
85° -- --
80° -- --
75° -- --
70° -- --
65° 5 02
60° 6 43
55° 7 47
50° 8 33
45° 9 09
40° 9 39
35° 10 04
30° 10 25
25° 10 45
GULF OF MEXICO
ALASKA
HAWAII
Fairbanks
Anchorage
Honolulu
Williston
International Falls
Rapid City
North Platte
Des Moines
Dodge City
Amarillo
Oklahoma City
Little Rock
Fort Worth
Shreveport
New Orleans
Houston
San Antonio
Del Rio
Brownsville
El Paso
Albuquerque
Phoenix
Las Vegas
Los Angeles
San Diego
Yuma
Fresno
San Francisco
Sacramento
Reno
Winnemucca
Eureka
Roseburg
Portland
Pocatello
Salt Lake City
Grand Junction
Cheyenne
Denver
Havre
Great Falls
Billings
Bismarck
Fargo
Huron
Lincoln
Kansas City
Springfield
St. Louis
Minneapolis
Duluth
Green Bay
Chicago
Dubuque
Lansing
Detroit
Sault Ste. Marie
Buffalo
Pittsburgh
Harrisburg
Philadelphia
New York
Boston
Albany
Burlington
Concord
Portland
Caribou
Columbus
Indianapolis
Louisville
Lexington
Charleston
Washington, D.C.
Richmond
Norfolk
Raleigh
Knoxville
Nashville
Memphis
Birmingham
Atlanta
Columbia
Charleston
Wilmington
Jacksonville
Pensacola
Jackson
Tampa
Miami

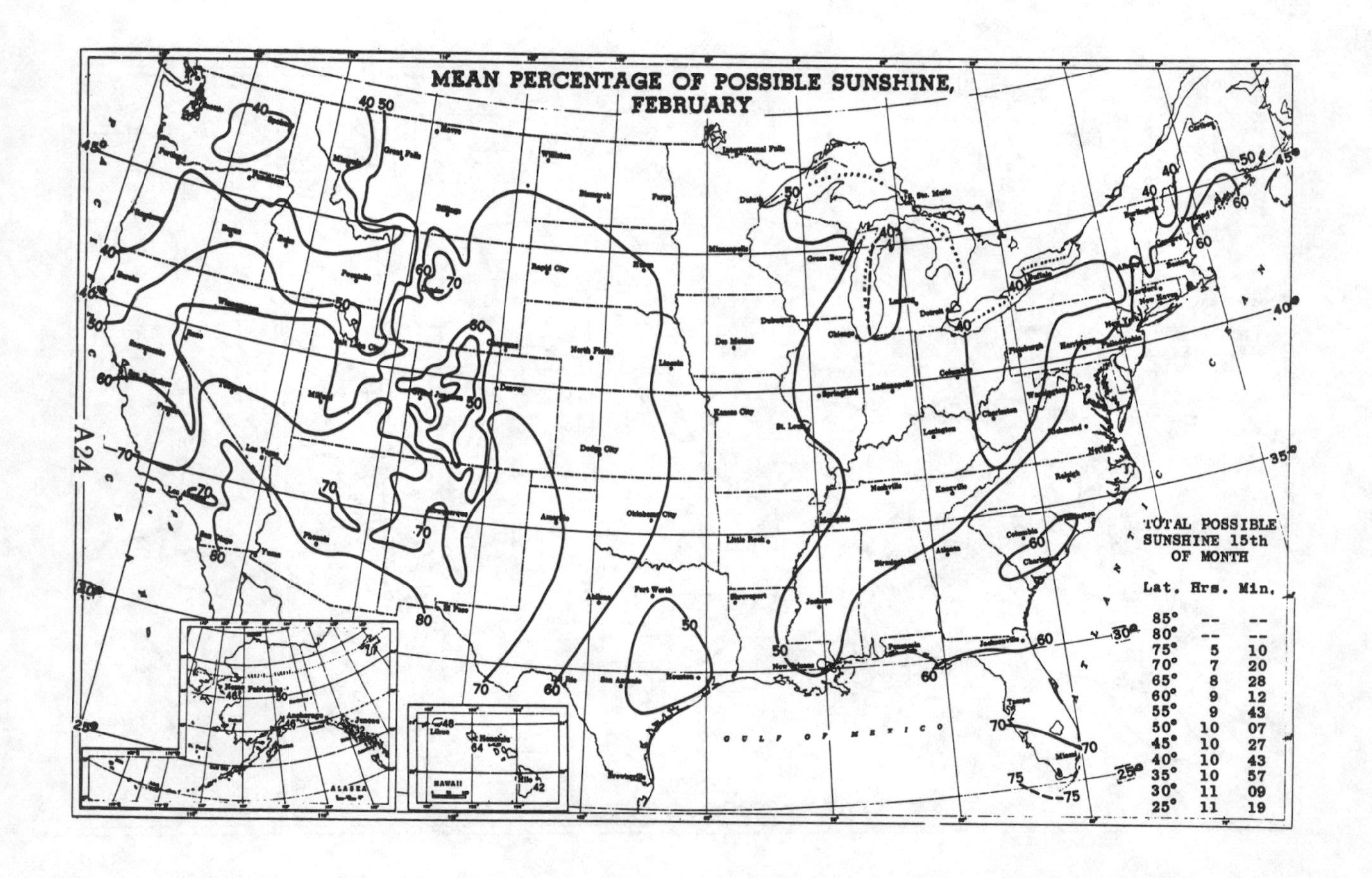

Lat.	Hrs.	Min.
85°	--	--
80°	--	--
75°	5	10
70°	7	20
65°	8	28
60°	9	12
55°	9	43
50°	10	07
45°	10	27
40°	10	43
35°	10	57
30°	11	09
25°	11	19

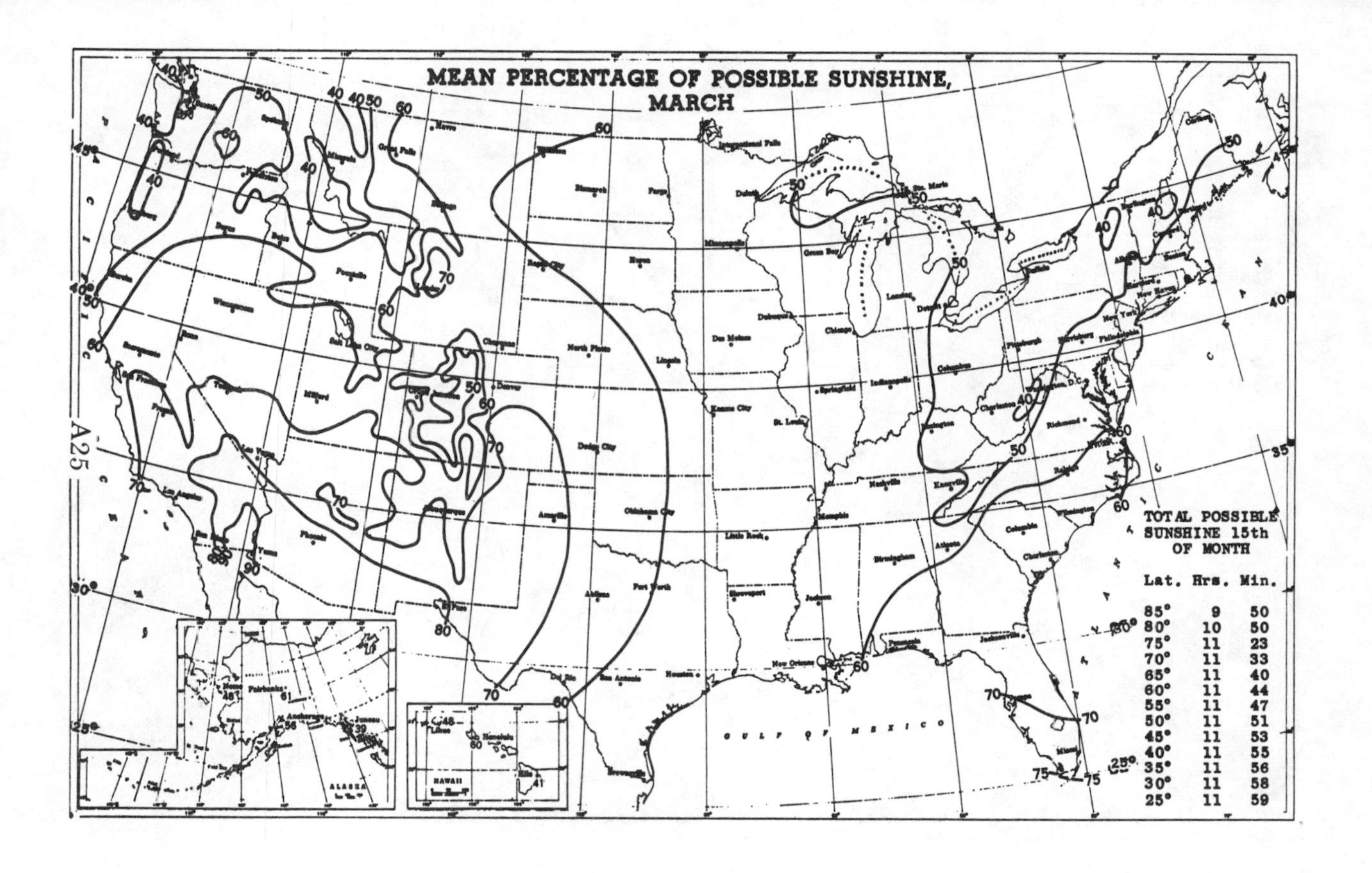
MEAN PERCENTAGE OF POSSIBLE SUNSHINE,
MARCH
TOTAL POSSIBLE
SUNSHINE 15th
OF MONTH
Lat. Hrs. Min.
85° 9 50
80° 10 50
75° 11 23
70° 11 33
65° 11 40
60° 11 44
55° 11 47
50° 11 51
45° 11 53
40° 11 55
35° 11 56
30° 11 58
25° 11 59
GULF OF MEXICO
ALASKA
HAWAII

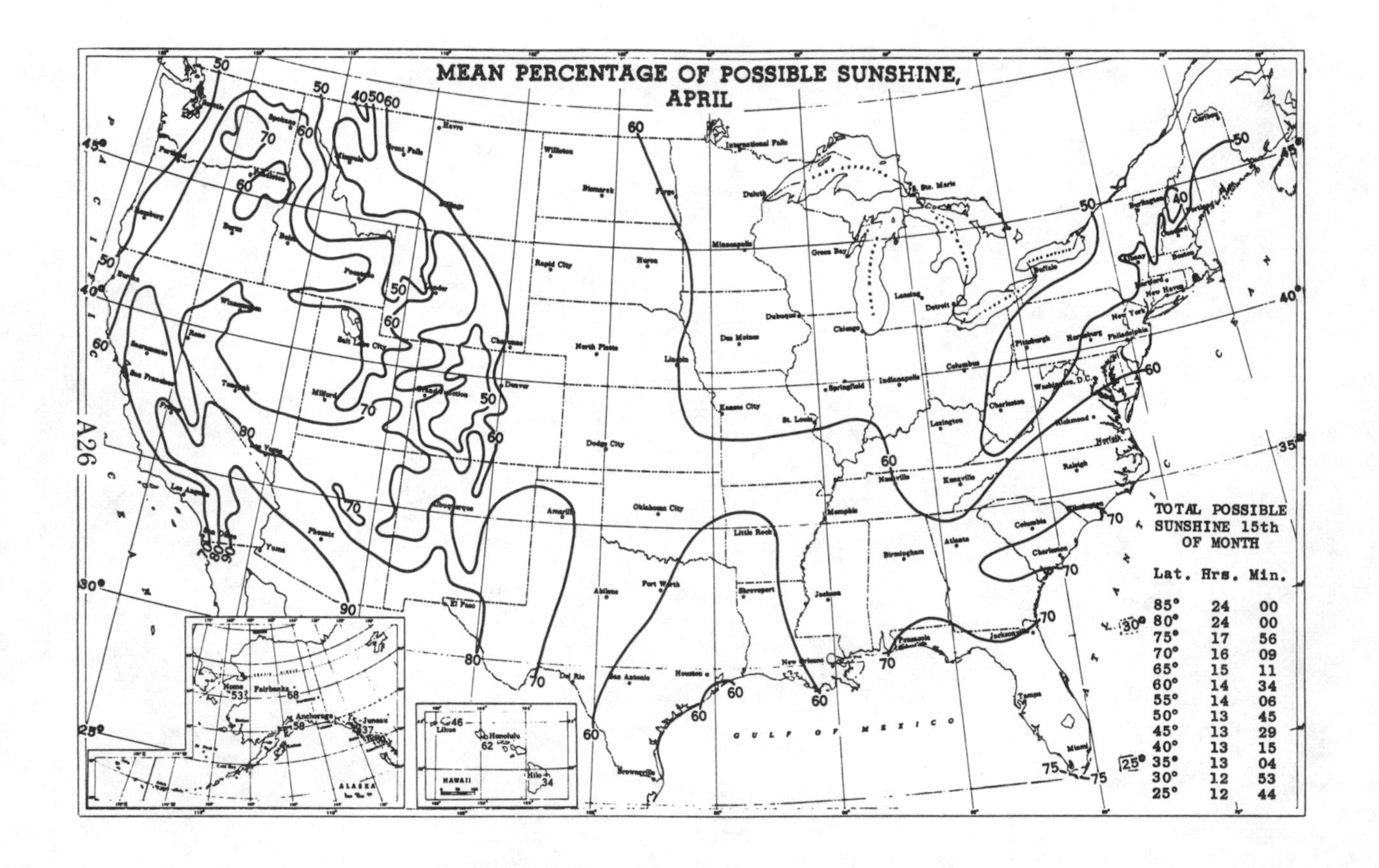

TOTAL POSSIBLE SUNSHINE 15th OF MONTH

Lat.	Hrs.	Min.
85°	24	00
80°	24	00
75°	17	56
70°	16	09
65°	15	11
60°	14	34
55°	14	06
50°	13	45
45°	13	29
40°	13	15
35°	13	04
30°	12	53
25°	12	44

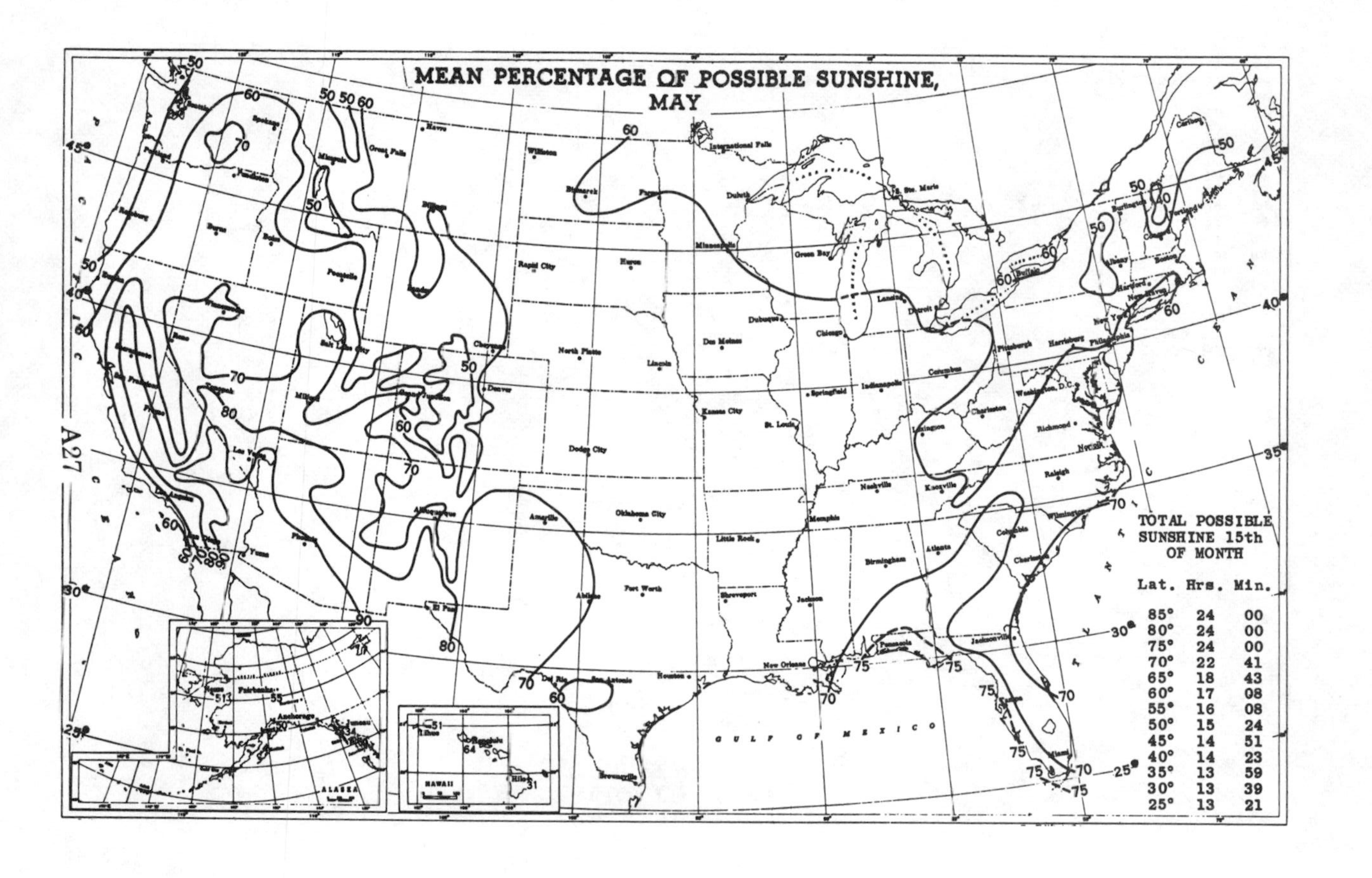

MEAN PERCENTAGE OF POSSIBLE SUNSHINE,
MAY
TOTAL POSSIBLE
SUNSHINE 15th
OF MONTH
Lat. Hrs. Min.
85° 24 00
80° 24 00
75° 24 00
70° 22 41
65° 18 43
60° 17 08
55° 16 08
50° 15 24
45° 14 51
40° 14 23
35° 13 59
30° 13 39
25° 13 21
GULF OF MEXICO
HAWAII
ALASKA

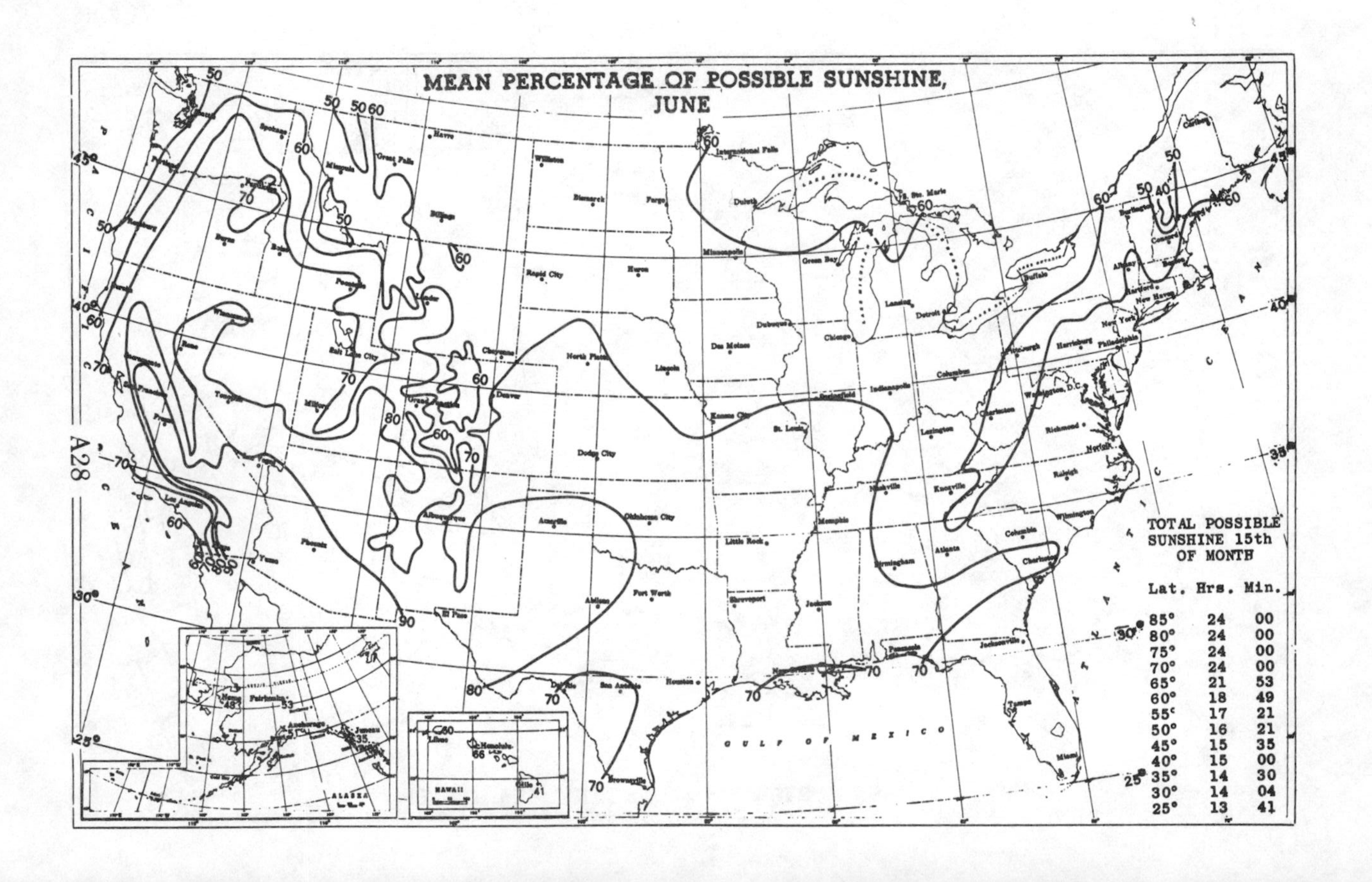
MEAN PERCENTAGE OF POSSIBLE SUNSHINE,
JUNE
TOTAL POSSIBLE
SUNSHINE 15th
OF MONTH
Lat. Hrs. Min.
85° 24 00
80° 24 00
75° 24 00
70° 24 00
65° 21 53
60° 18 49
55° 17 21
50° 16 21
45° 15 35
40° 15 00
35° 14 30
30° 14 04
25° 13 41
GULF OF MEXICO
HAWAII
ALASKA

MEAN PERCENTAGE OF POSSIBLE SUNSHINE, JULY

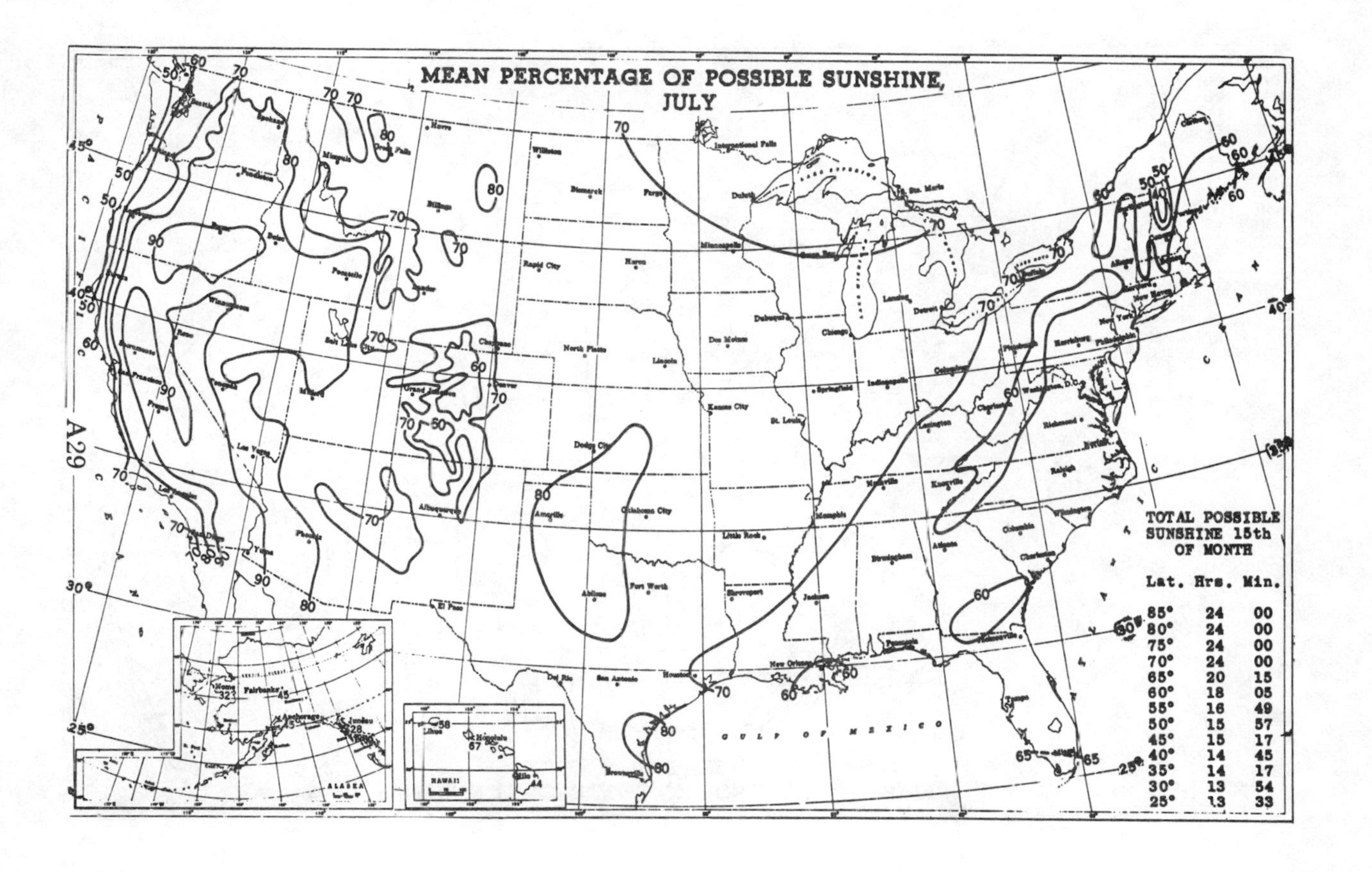

TOTAL POSSIBLE SUNSHINE 15th OF MONTH

Lat.	Hrs.	Min.
85°	24	00
80°	24	00
75°	24	00
70°	24	00
65°	20	15
60°	18	05
55°	16	49
50°	15	57
45°	15	17
40°	14	45
35°	14	17
30°	13	54
25°	13	33

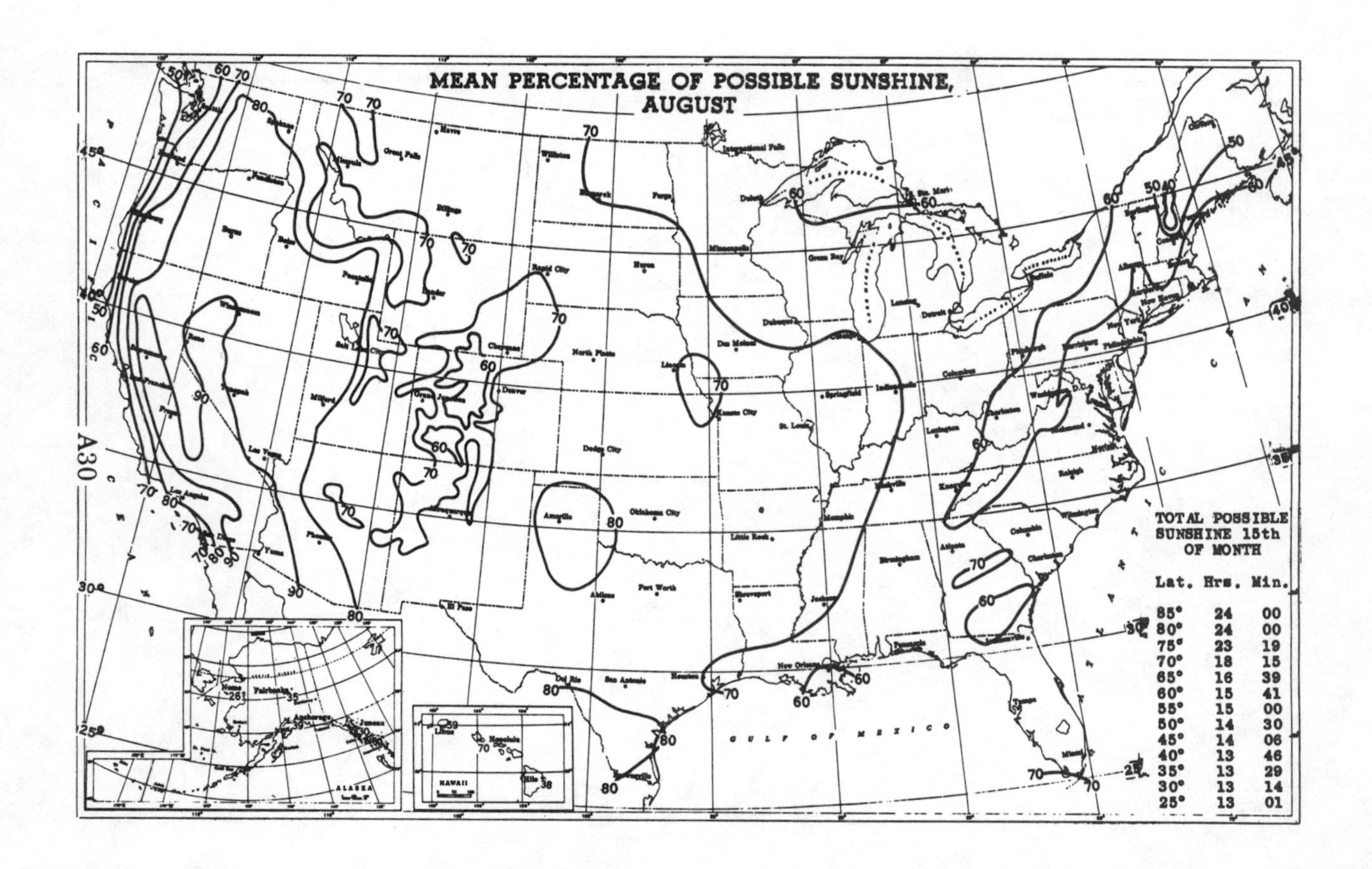
MEAN PERCENTAGE OF POSSIBLE SUNSHINE,
AUGUST
TOTAL POSSIBLE
SUNSHINE 15th
OF MONTH
Lat. Hrs. Min.
85° 24 00
80° 24 00
75° 23 19
70° 18 15
65° 16 39
60° 15 41
55° 15 00
50° 14 30
45° 14 06
40° 13 46
35° 13 29
30° 13 14
25° 13 01
GULF OF MEXICO
HAWAII
ALASKA

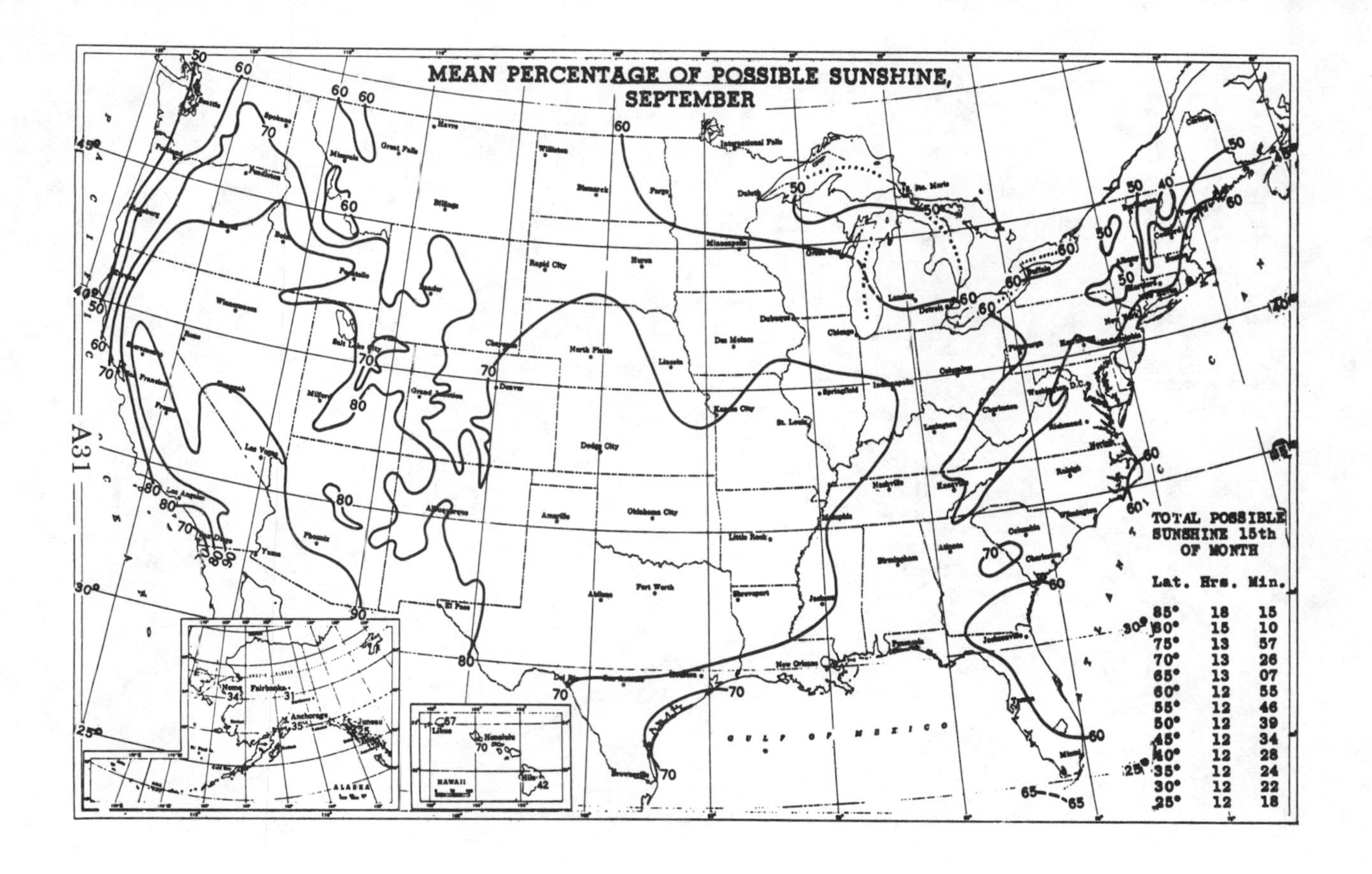

Lat.	Hrs.	Min.
85°	18	15
80°	15	10
75°	13	57
70°	13	26
65°	13	07
60°	12	55
55°	12	46
50°	12	39
45°	12	34
40°	12	28
35°	12	24
30°	12	22
25°	12	18

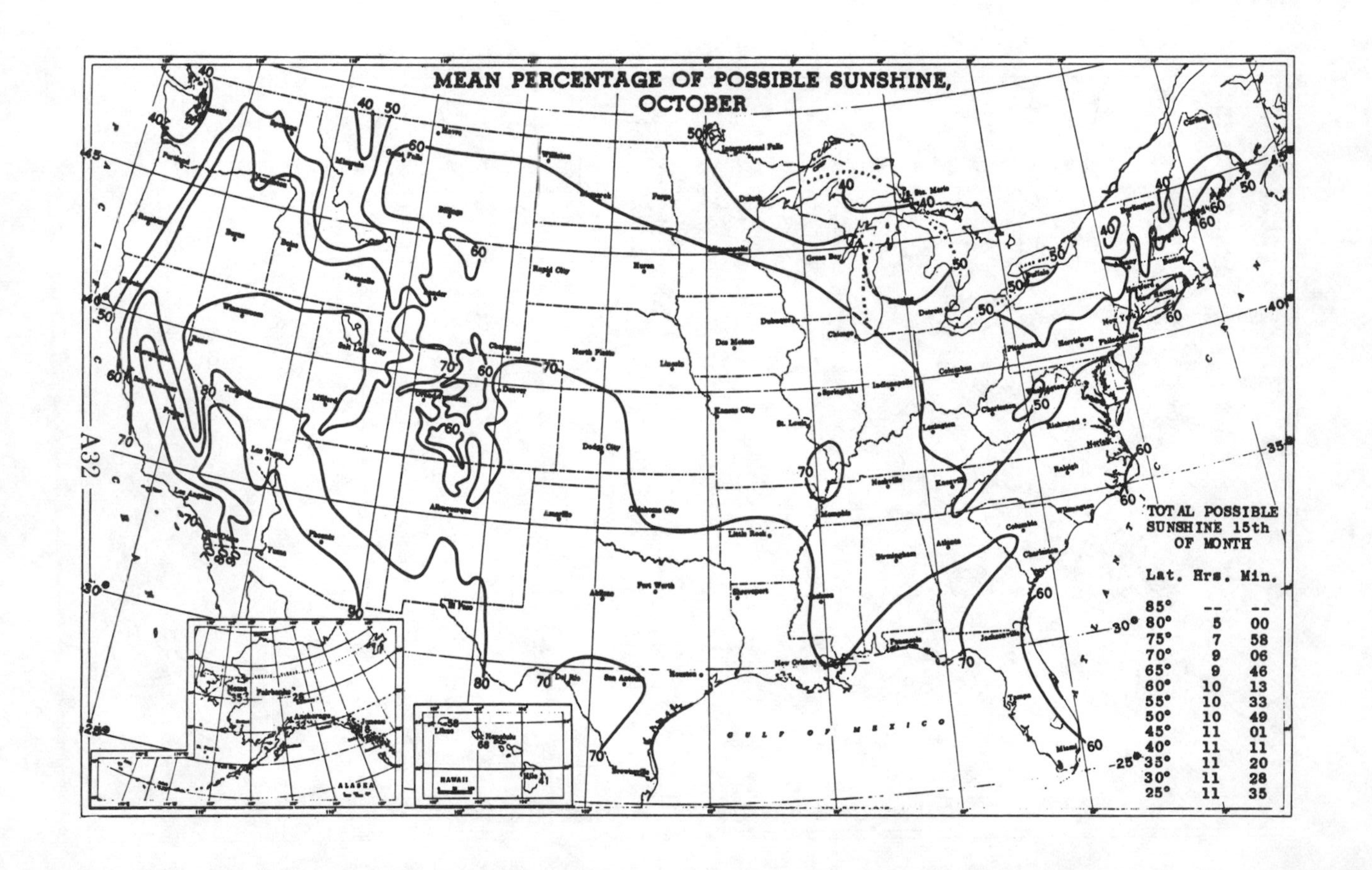
MEAN PERCENTAGE OF POSSIBLE SUNSHINE, OCTOBER
TOTAL POSSIBLE SUNSHINE 15th OF MONTH
Lat. Hrs. Min.
85° -- --
80° 5 00
75° 7 58
70° 9 06
65° 9 46
60° 10 13
55° 10 33
50° 10 49
45° 11 01
40° 11 11
35° 11 20
30° 11 28
25° 11 35
GULF OF MEXICO
HAWAII
ALASKA

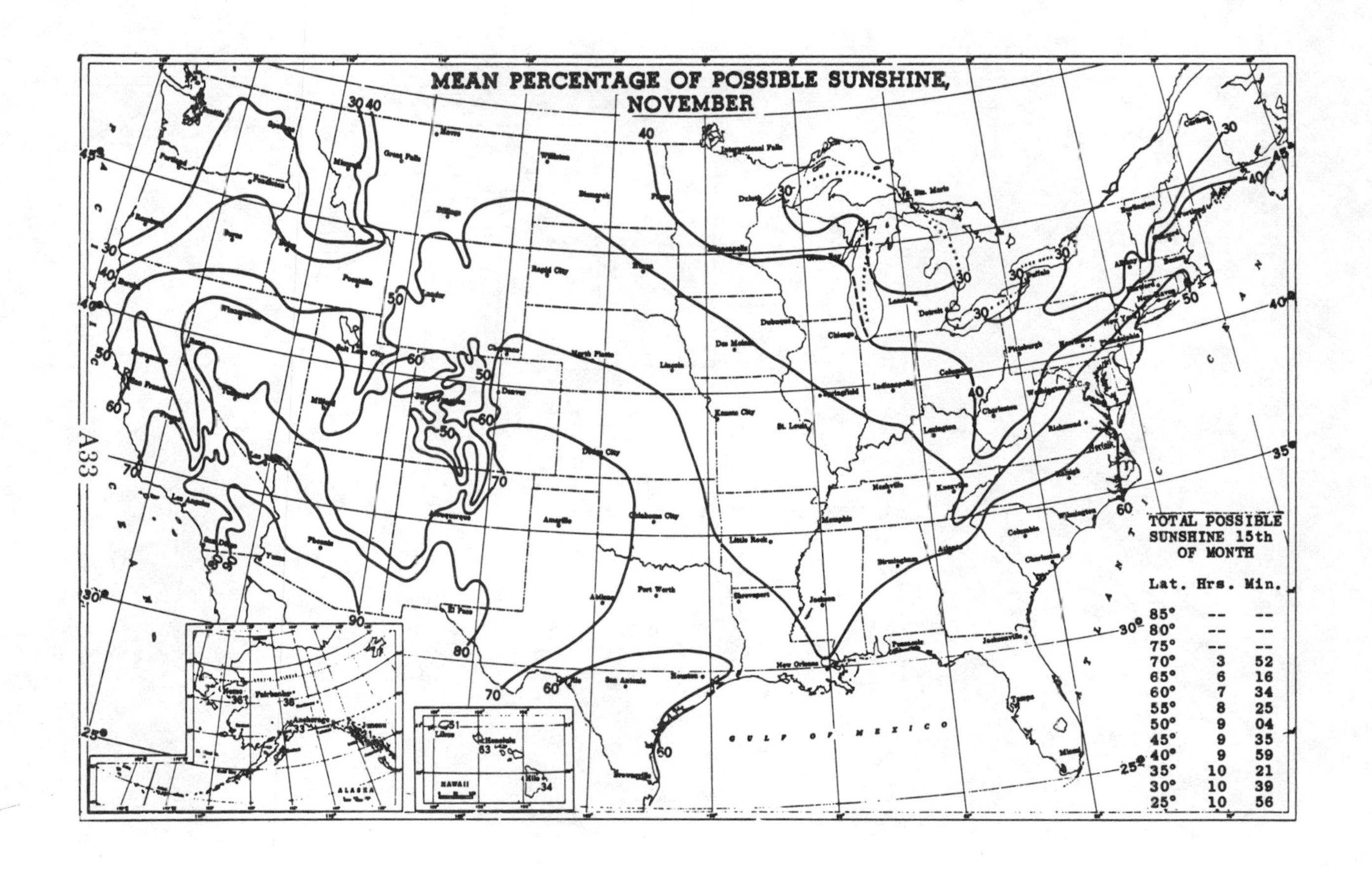

MEAN PERCENTAGE OF POSSIBLE SUNSHINE,
NOVEMBER
TOTAL POSSIBLE
SUNSHINE 15th
OF MONTH
Lat. Hrs. Min.
85° -- --
80° -- --
75° -- --
70° 3 52
65° 6 16
60° 7 34
55° 8 25
50° 9 04
45° 9 35
40° 9 59
35° 10 21
30° 10 39
25° 10 56
GULF OF MEXICO
ALASKA
HAWAII

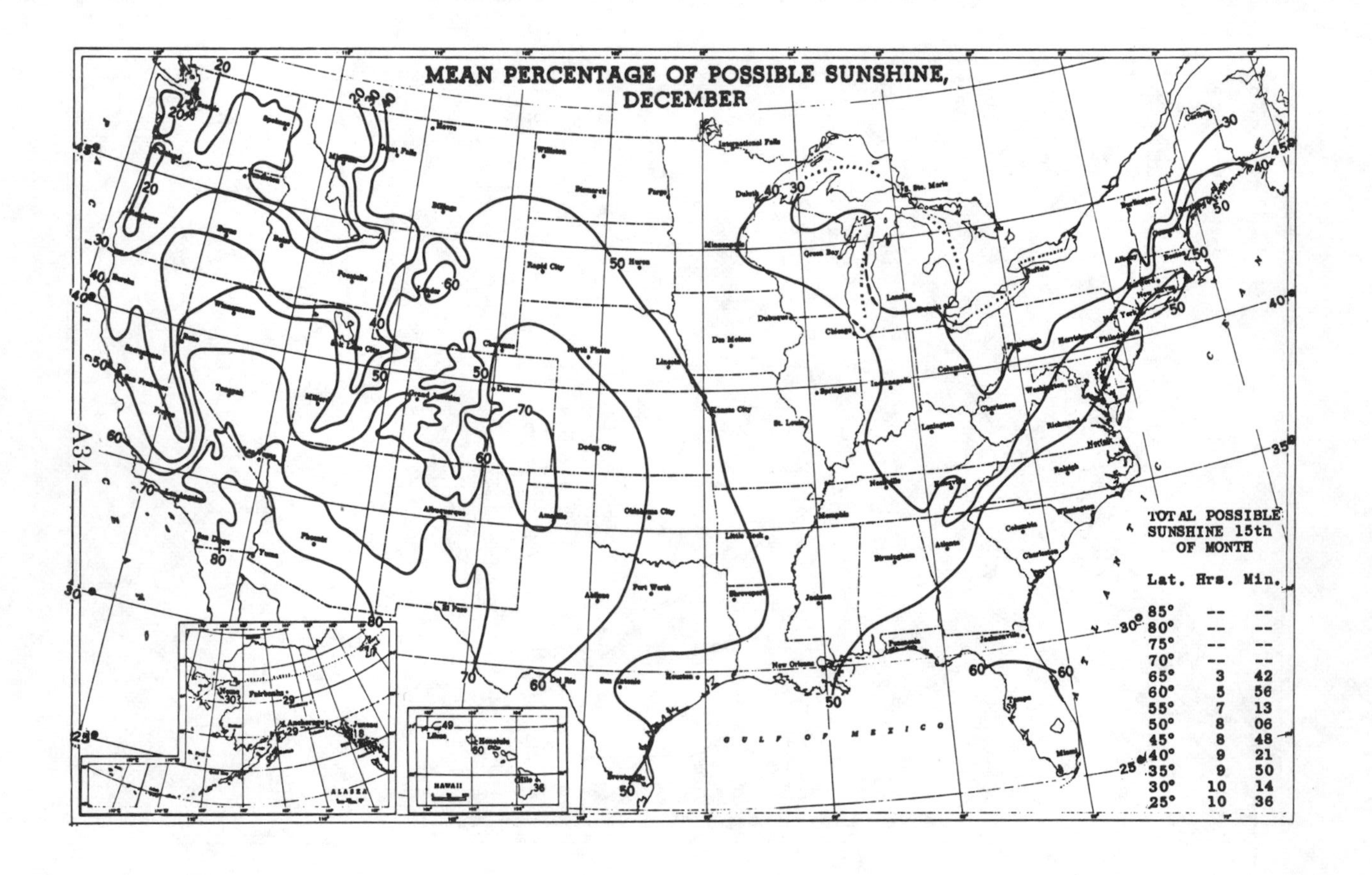

TOTAL POSSIBLE SUNSHINE 15th OF MONTH

Lat.	Hrs.	Min.
85°	--	--
80°	--	--
75°	--	--
70°	--	--
65°	3	42
60°	5	56
55°	7	13
50°	8	06
45°	8	48
40°	9	21
35°	9	50
30°	10	14
25°	10	36

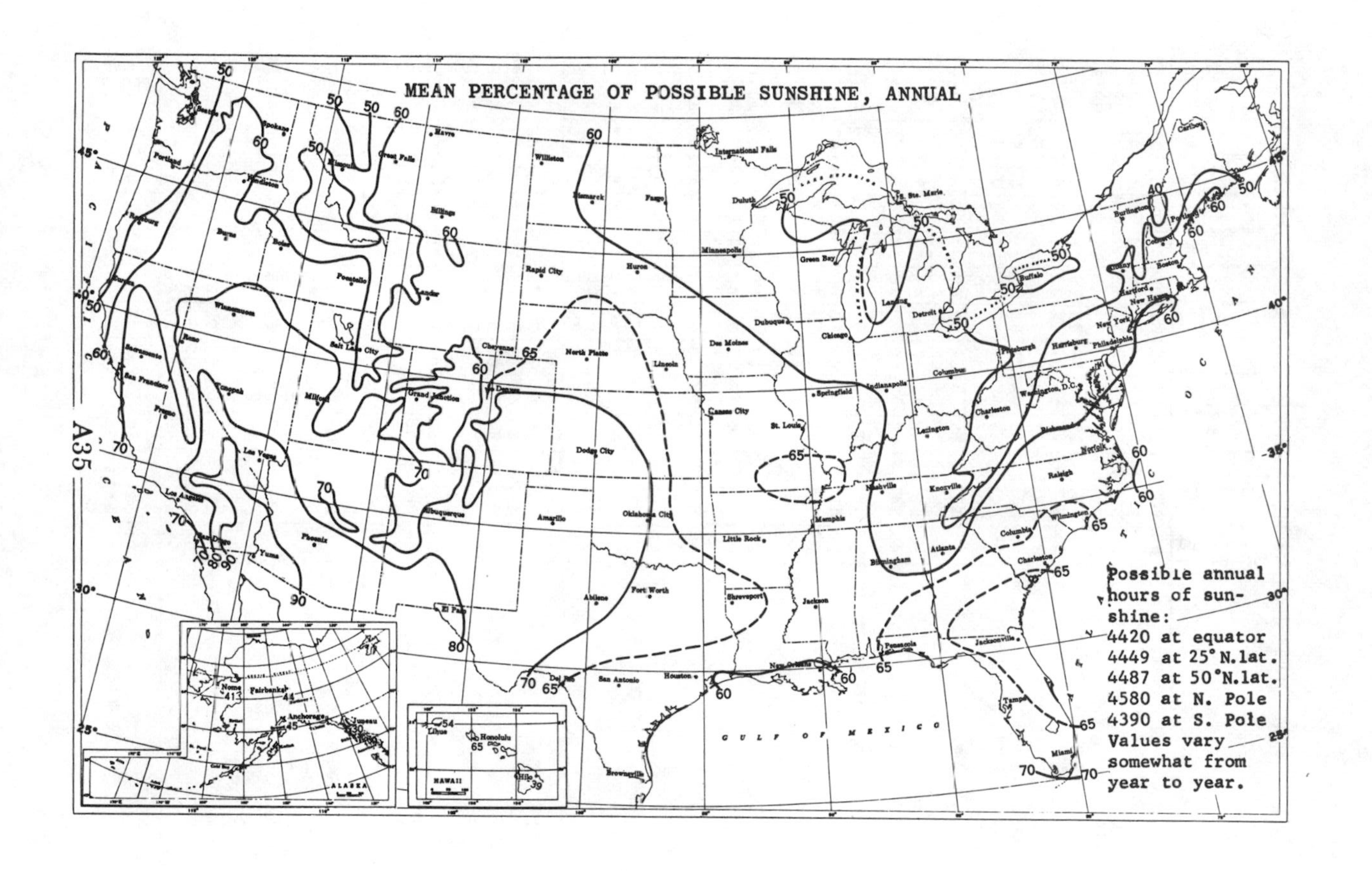
MEAN PERCENTAGE OF POSSIBLE SUNSHINE, ANNUAL
Possible annual
hours of sun-
shine:
4420 at equator
4449 at 25°N.lat.
4487 at 50°N.lat.
4580 at N. Pole
4390 at S. Pole
Values vary
somewhat from
year to year.
GULF OF MEXICO
ALASKA
HAWAII

MEAN PERCENTAGE OF POSSIBLE SUNSHINE

STATE AND STATION	YEARS	JAN.	FEB.	MAR.	APR.	MAY	JUNE	JULY	AUG.	SEPT.	OCT.	NOV.	DEC.	ANNUAL
ALA. BIRMINGHAM	56	43	49	56	63	66	67	62	65	66	67	58	44	59
MONTGOMERY	49	51	53	61	69	73	72	66	69	69	71	64	48	64
ALASKA. ANCHORAGE	19	39	46	56	58	50	51	45	39	35	32	33	29	45
FAIRBANKS	20	34	50	61	68	55	53	45	35	31	28	38	29	44
JUNEAU	14	30	32	39	37	34	35	28	30	25	18	21	18	30
NOME	29	44	46	48	53	51	48	32	26	34	35	36	30	41
ARIZ. PHOENIX	64	76	79	83	88	93	94	84	84	89	88	84	77	85
YUMA	52	83	87	91	94	97	98	92	91	93	93	90	83	91
ARK. LITTLE ROCK	66	44	53	57	62	67	72	71	73	71	74	58	47	62
CALIF. EUREKA	49	40	44	50	53	54	56	51	46	52	48	42	39	49
FRESNO	55	46	63	72	83	89	94	97	97	93	87	73	47	78
LOS ANGELES	63	70	69	70	67	68	69	80	81	80	76	79	72	73
RED BLUFF	39	50	60	65	75	79	86	95	94	89	77	64	50	75
SACRAMENTO	48	44	57	67	76	82	90	96	95	92	82	65	44	77
SAN DIEGO	68	68	67	68	66	60	60	67	70	70	70	76	71	68
SAN FRANCISCO	64	53	57	63	69	70	75	68	63	70	70	62	54	66
COLO. DENVER	64	67	67	65	63	61	69	68	68	71	71	67	65	67
GRAND JUNCTION	57	58	62	64	67	71	79	76	72	77	74	67	58	69
CONN. HARTFORD	48	46	55	56	54	57	60	62	60	57	55	46	46	56
D. C. WASHINGTON	66	46	53	56	57	61	64	64	62	62	61	54	47	58
FLA. APALACHICOLA	26	59	62	62	71	77	70	64	63	62	74	66	53	65
JACKSONVILLE	60	58	59	66	71	71	63	62	63	58	58	61	53	62
KEY WEST	45	68	75	78	78	76	70	69	71	65	65	69	66	71
MIAMI BEACH	48	66	72	73	73	68	62	65	67	62	62	65	65	67
TAMPA	63	63	67	71	74	75	66	61	64	64	67	67	61	68
GA. ATLANTA	65	48	53	57	65	68	68	62	63	65	67	60	47	60
HAWAII. HILO	9	48	42	41	34	31	41	44	38	42	41	34	36	39
HONOLULU	53	62	64	60	62	64	66	67	70	70	68	63	60	65
LIHUE	9	48	48	48	46	51	60	58	59	67	58	51	49	54
IDAHO. BOISE	20	40	48	59	67	68	75	89	86	81	66	46	37	66
POCATELLO	21	37	47	58	64	66	72	82	81	78	66	48	36	64
ILL. CAIRO	30	46	53	59	65	71	77	82	79	75	73	56	46	65
CHICAGO	66	44	49	53	56	63	69	73	70	65	61	47	41	59
SPRINGFIELD	59	47	51	54	58	64	69	76	72	73	64	53	45	60
IND. EVANSVILLE	48	42	49	55	61	67	73	78	76	73	67	52	42	64
FT. WAYNE	48	38	44	51	55	62	69	74	69	64	58	41	38	57
INDIANAPOLIS	63	41	47	49	55	62	68	74	70	68	64	48	39	59
IOWA. DES MOINES	66	56	56	56	59	62	66	75	70	64	64	53	48	62
DUBUQUE	54	48	52	52	58	60	63	73	67	61	55	44	40	57
SIOUX CITY	52	55	58	58	59	63	67	75	72	67	65	53	50	63
KANS. CONCORDIA	52	60	60	62	63	65	73	79	76	72	70	64	58	67
DODGE CITY	70	67	66	68	68	68	74	78	78	76	75	70	67	71
WICHITA	46	61	63	64	64	66	73	80	77	73	69	67	59	69
KY. LOUISVILLE	59	41	47	52	57	64	68	72	69	68	64	51	39	59
LA. NEW ORLEANS	69	49	50	57	63	66	64	58	60	64	70	60	46	59
SHREVEPORT	18	48	54	58	60	69	78	79	80	79	77	65	60	69
MAINE. EASTPORT	58	45	51	52	52	51	53	55	57	54	50	37	40	50
MASS. BOSTON	67	47	56	57	56	59	62	64	63	61	58	48	48	57
MICH. ALPENA	45	29	43	52	56	59	64	70	64	52	44	24	22	51
DETROIT	69	34	42	48	52	58	65	69	66	61	54	35	29	53
GRAND RAPIDS	56	26	37	48	54	60	66	72	67	58	50	31	22	49
MARQUETTE	55	31	40	47	52	53	56	63	57	47	38	24	24	47
S. STE. MARIE	60	28	44	50	54	54	59	63	58	45	36	21	22	47

MEAN PERCENTAGE OF POSSIBLE SUNSHINE

STATE AND STATION	YEARS	JAN.	FEB.	MAR.	APR.	MAY	JUNE	JULY	AUG.	SEPT.	OCT.	NOV.	DEC.	ANNUAL
MINN. DULUTH	49	47	55	60	58	58	60	68	63	53	47	36	40	55
MINNEAPOLIS	45	49	54	55	57	60	64	72	69	60	54	40	40	56
MISS. VICKSBURG	66	46	50	57	64	69	73	69	72	74	71	60	45	64
MO. KANSAS CITY	69	55	57	59	60	64	70	76	73	70	67	59	52	65
ST. LOUIS	68	48	49	56	59	64	68	72	68	67	65	54	44	61
SPRINGFIELD	45	48	54	57	60	63	69	77	72	71	65	58	48	63
MONT. HAVRE	55	49	58	61	63	63	65	78	75	64	57	48	46	62
HELENA	65	46	55	58	60	59	63	77	74	63	57	48	43	60
KALISPELL	50	28	40	49	57	58	60	77	73	61	50	28	20	53
NEBR. LINCOLN	55	57	59	60	60	63	69	76	71	67	66	59	55	64
NORTH PLATTE	53	63	63	64	62	64	72	78	74	72	70	62	58	68
NEV. ELY	21	61	64	68	65	67	79	79	81	81	73	67	62	72
LAS VEGAS	19	74	77	78	81	85	91	84	86	92	84	83	75	82
RENO	51	59	64	69	75	77	82	90	89	86	76	68	56	76
WINNEMUCCA	53	52	60	64	70	76	83	90	90	86	75	62	53	74
N. H. CONCORD	44	48	53	55	53	51	56	57	58	55	50	43	43	52
N. J. ATLANTIC CITY	62	51	57	58	59	62	65	67	66	65	54	58	52	60
N. MEX. ALBUQUERQUE	28	70	72	72	76	79	84	76	75	81	80	79	70	76
ROSWELL	47	69	72	75	77	76	80	76	75	74	74	74	69	74
N. Y. ALBANY	63	43	51	53	53	57	62	63	61	58	54	39	38	53
BINGHAMTON	63	31	39	41	44	50	56	54	51	47	43	29	26	44
BUFFALO	49	32	41	49	51	59	67	70	67	60	51	31	28	53
CANTON	43	37	47	50	48	54	61	63	61	54	45	30	31	49
NEW YORK	83	49	56	57	59	62	65	66	64	64	61	53	50	59
SYRACUSE	49	31	38	45	50	58	64	67	63	56	47	29	26	50
N. C. ASHEVILLE	57	48	53	56	61	64	63	59	59	62	64	59	48	58
RALEIGH	61	50	56	59	64	67	65	62	62	63	64	62	52	61
N. DAK. BISMARCK	65	52	58	56	57	58	61	73	69	62	59	49	48	59
DEVILS LAKE	55	53	60	59	60	59	62	71	67	59	56	44	45	58
FARGO	39	47	55	56	58	62	63	73	69	60	57	39	46	59
WILLISTON	43	51	59	60	63	66	66	78	75	65	60	48	48	63
OHIO. CINCINNATI	44	41	46	52	56	62	69	72	68	68	60	46	39	57
CLEVELAND	65	29	36	45	52	61	67	71	68	62	54	32	25	50
COLUMBUS	65	36	44	49	54	63	68	71	68	66	60	44	35	55
OKLA. OKLAHOMA CITY	62	57	60	63	64	65	74	78	78	74	68	64	57	68
OREG. BAKER	46	41	49	56	61	63	67	83	81	74	62	46	37	60
PORTLAND	69	27	34	41	49	52	55	70	65	55	42	28	23	48
ROSEBURG	29	24	32	40	51	57	59	79	77	68	42	28	18	51
PA. HARRISBURG	60	43	52	55	57	61	65	68	63	62	58	47	43	57
PHILADELPHIA	66	45	56	57	58	61	62	64	61	62	61	53	49	57
PITTSBURGH	63	32	39	45	50	57	62	64	61	62	54	39	30	51
R. I. BLOCK ISLAND	48	45	54	47	56	58	60	62	62	60	59	50	44	56
S. C. CHARLESTON	61	58	60	65	72	73	70	66	66	67	68	68	57	66
COLUMBIA	55	53	57	62	68	69	68	63	65	64	68	64	51	63
S. DAK. HURON	62	55	62	60	62	65	68	76	72	66	61	52	49	63
RAPID CITY	53	58	62	63	62	61	66	73	73	69	66	58	54	64
TENN. KNOXVILLE	62	42	49	53	59	64	66	64	59	64	64	53	41	57
MEMPHIS	55	44	51	57	64	68	74	73	74	70	69	58	45	64
NASHVILLE	63	42	47	54	60	65	69	69	68	69	65	55	42	59
TEX. ABILENE	14	64	68	73	66	73	86	83	85	73	71	72	66	73
AMARILLO	54	71	71	75	75	75	82	81	81	79	76	76	70	76
AUSTIN	33	46	50	57	60	62	72	76	79	70	70	57	49	63

MEAN PERCENTAGE OF POSSIBLE SUNSHINE

STATE AND STATION	YEARS	JAN.	FEB.	MAR.	APR.	MAY	JUNE	JULY	AUG.	SEPT.	OCT.	NOV.	DEC.	ANNUAL
BROWNSVILLE	37	44	49	51	57	65	73	78	78	67	70	54	44	61
DEL RIO	36	53	55	61	63	60	66	75	80	69	66	58	52	63
EL PASO	53	74	77	81	85	87	87	78	78	80	82	80	73	80
FT. WORTH	33	56	57	65	66	67	75	78	78	74	70	63	58	68
GALVESTON	66	50	50	55	61	69	76	72	71	70	74	62	49	63
SAN ANTONIO	57	48	51	56	58	60	69	74	75	69	67	55	49	62
UTAH. SALT LAKE CITY	22	48	53	61	68	73	78	82	82	84	73	56	49	69
VT. BURLINGTON	54	34	43	48	47	53	59	62	59	51	43	25	24	46
VA. NORFOLK	60	50	57	60	63	67	66	66	66	63	64	60	51	62
RICHMOND	56	49	55	59	63	67	66	65	62	63	64	58	50	61
WASH. NORTH HEAD	44	28	37	42	48	48	48	50	46	48	41	31	27	41
SEATTLE	26	27	34	42	48	53	48	62	56	53	36	28	24	45
SPOKANE	62	26	41	53	63	64	68	82	79	68	53	28	22	58
TATOOSH ISLAND	49	26	36	39	45	47	46	48	44	47	38	26	23	40
WALLA WALLA	44	24	35	51	63	67	72	86	84	72	59	33	20	60
YAKIMA	18	34	49	62	70	72	74	86	86	74	61	38	29	65
W. VA. ELKINS	55	33	37	42	47	55	55	56	53	55	51	41	33	48
PARKERSBURG	62	30	36	42	49	56	60	63	60	60	53	37	29	48
WIS. GREEN BAY	57	44	51	55	56	58	64	70	65	58	52	40	40	55
MADISON	59	44	49	52	53	58	64	70	66	60	56	41	38	56
MILWAUKEE	59	44	48	53	56	60	65	73	67	62	56	44	39	57
WYO. CHEYENNE	63	65	66	64	61	59	68	70	68	69	69	65	63	66
LANDER	57	66	70	71	66	65	74	76	75	72	67	61	62	69
SHERIDAN	52	56	61	62	61	61	67	76	74	67	60	53	52	64
YELLOWSTONE PARK	35	39	51	55	57	56	63	73	71	65	57	45	38	56
P. R. SAN JUAN	57	64	69	71	66	59	62	65	67	61	63	63	65	65

Based on period of record through December 1959, except in a few instances.

These charts and tabulation derived from "Normals, Means, and Extremes" table in U. S. Weather Bureau publication **Local Climatological Data,** except inset table on charts from U. S. Naval Observatory publication **Tables of Sunrise, Sunset, and Twilight.**

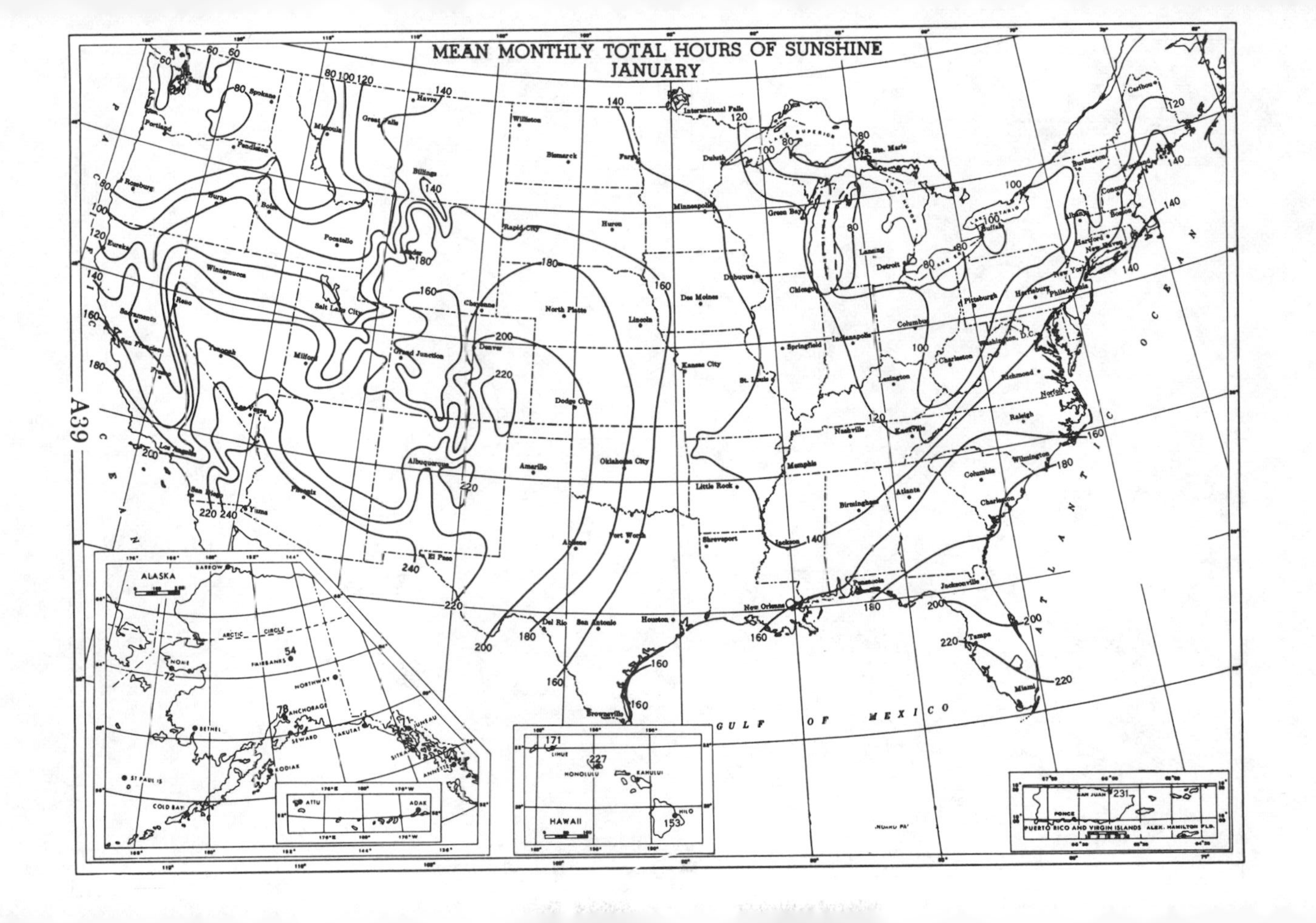
MEAN MONTHLY TOTAL HOURS OF SUNSHINE
JANUARY
ALASKA
HAWAII
PUERTO RICO AND VIRGIN ISLANDS
GULF OF MEXICO

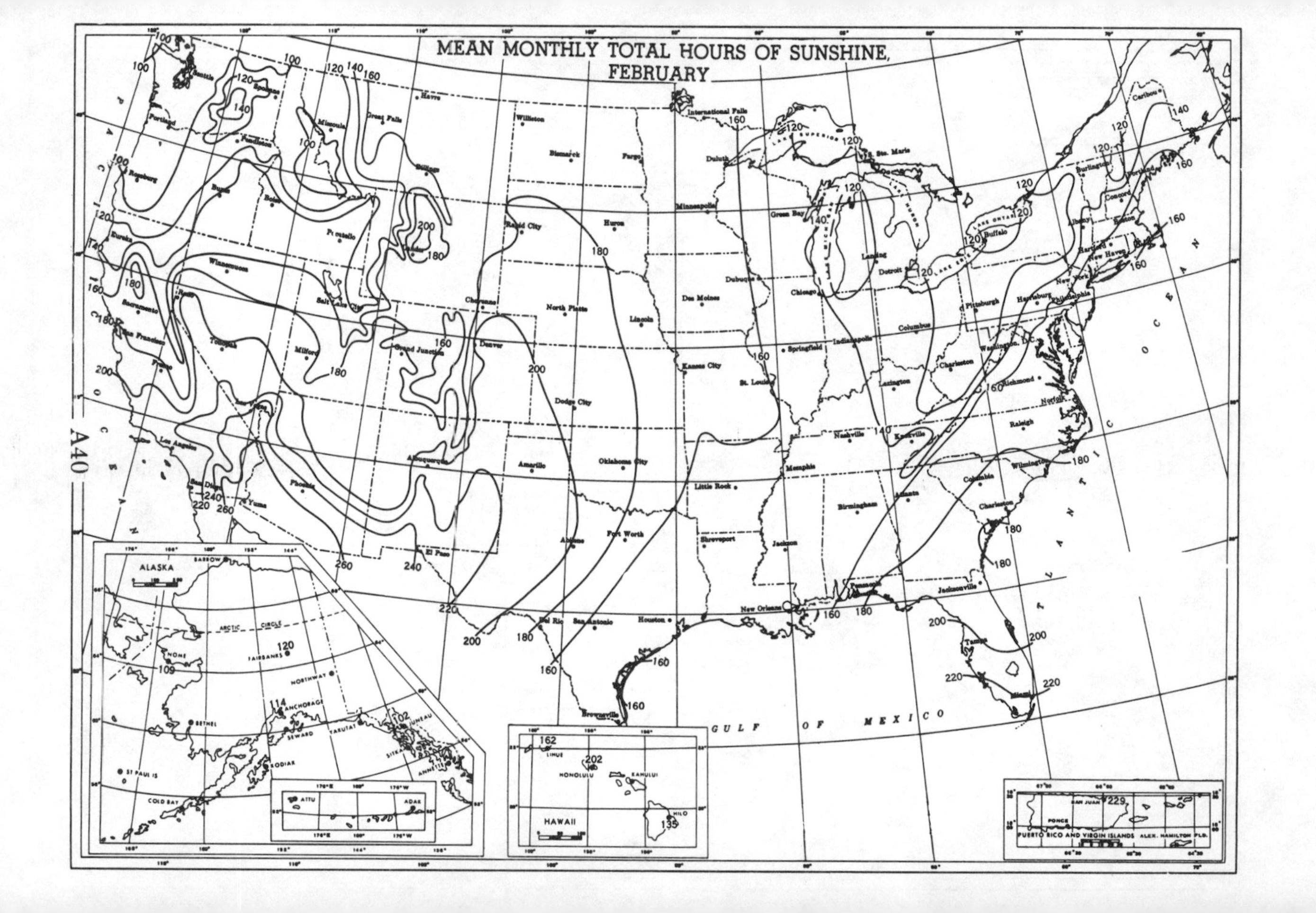
MEAN MONTHLY TOTAL HOURS OF SUNSHINE,
FEBRUARY
GULF OF MEXICO
ALASKA
HAWAII
PUERTO RICO AND VIRGIN ISLANDS

MEAN MONTHLY TOTAL HOURS OF SUNSHINE, MARCH

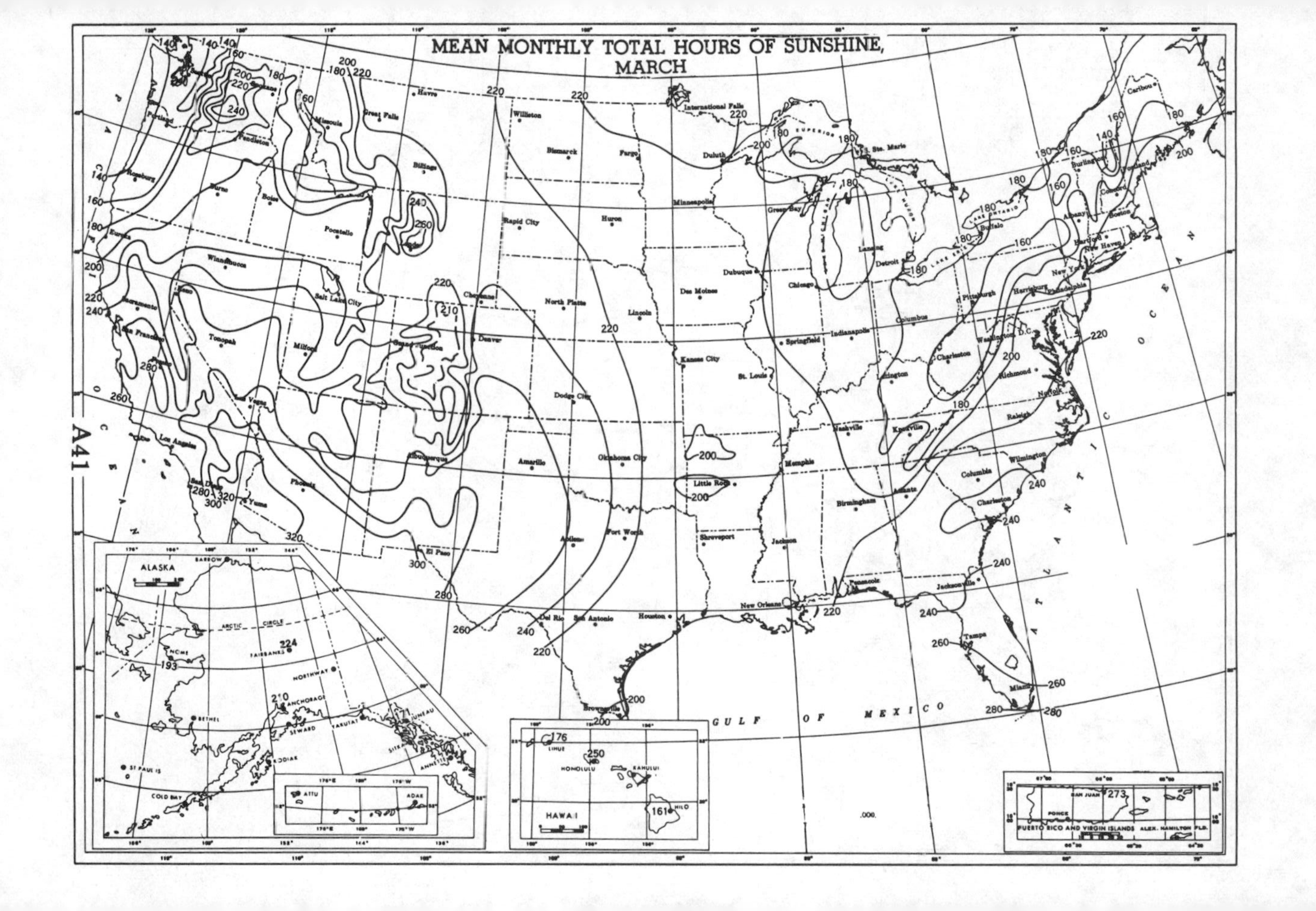

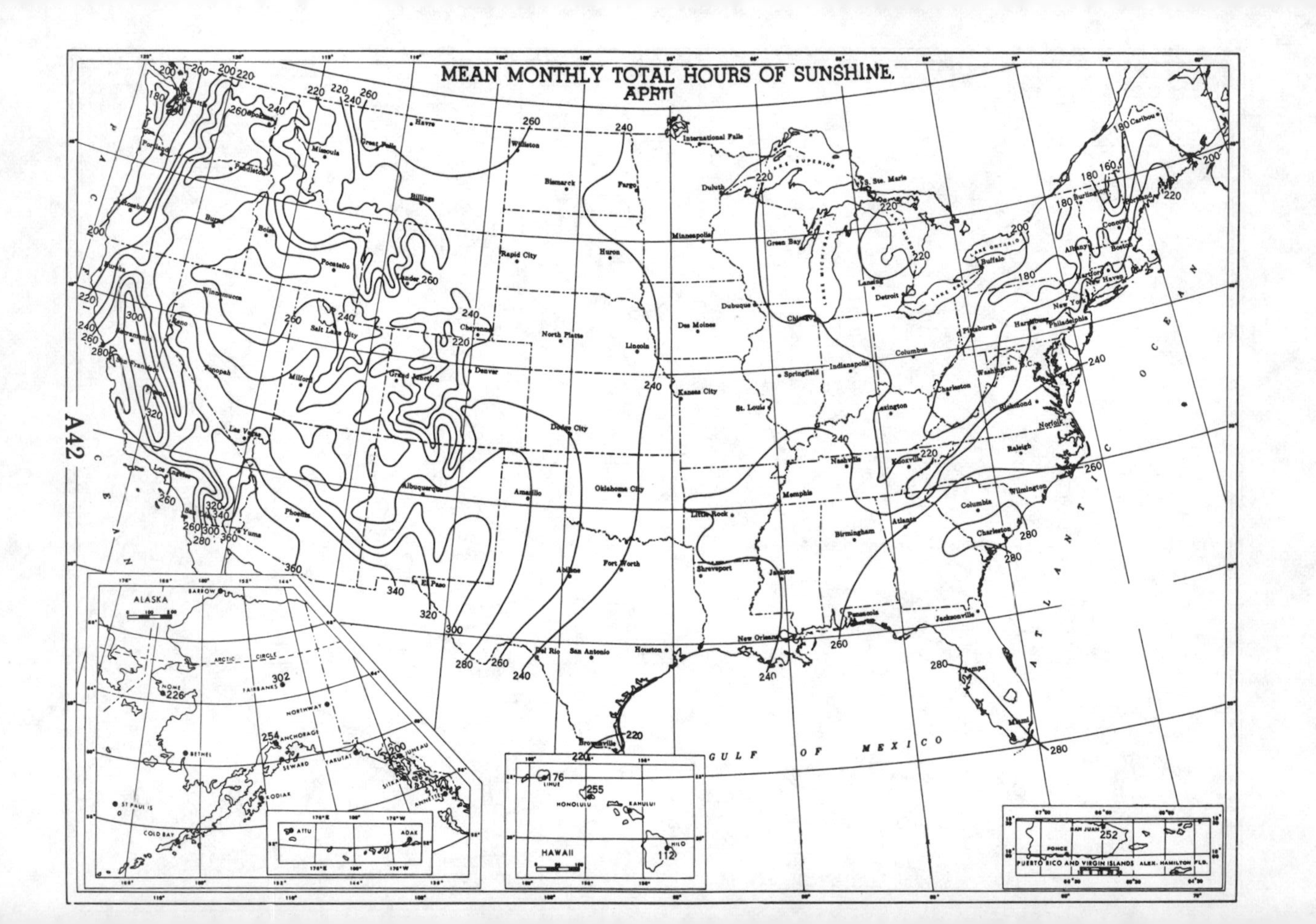
MEAN MONTHLY TOTAL HOURS OF SUNSHINE,
APRIL
ALASKA
HAWAII
PUERTO RICO AND VIRGIN ISLANDS
GULF OF MEXICO
PACIFIC OCEAN
ATLANTIC OCEAN

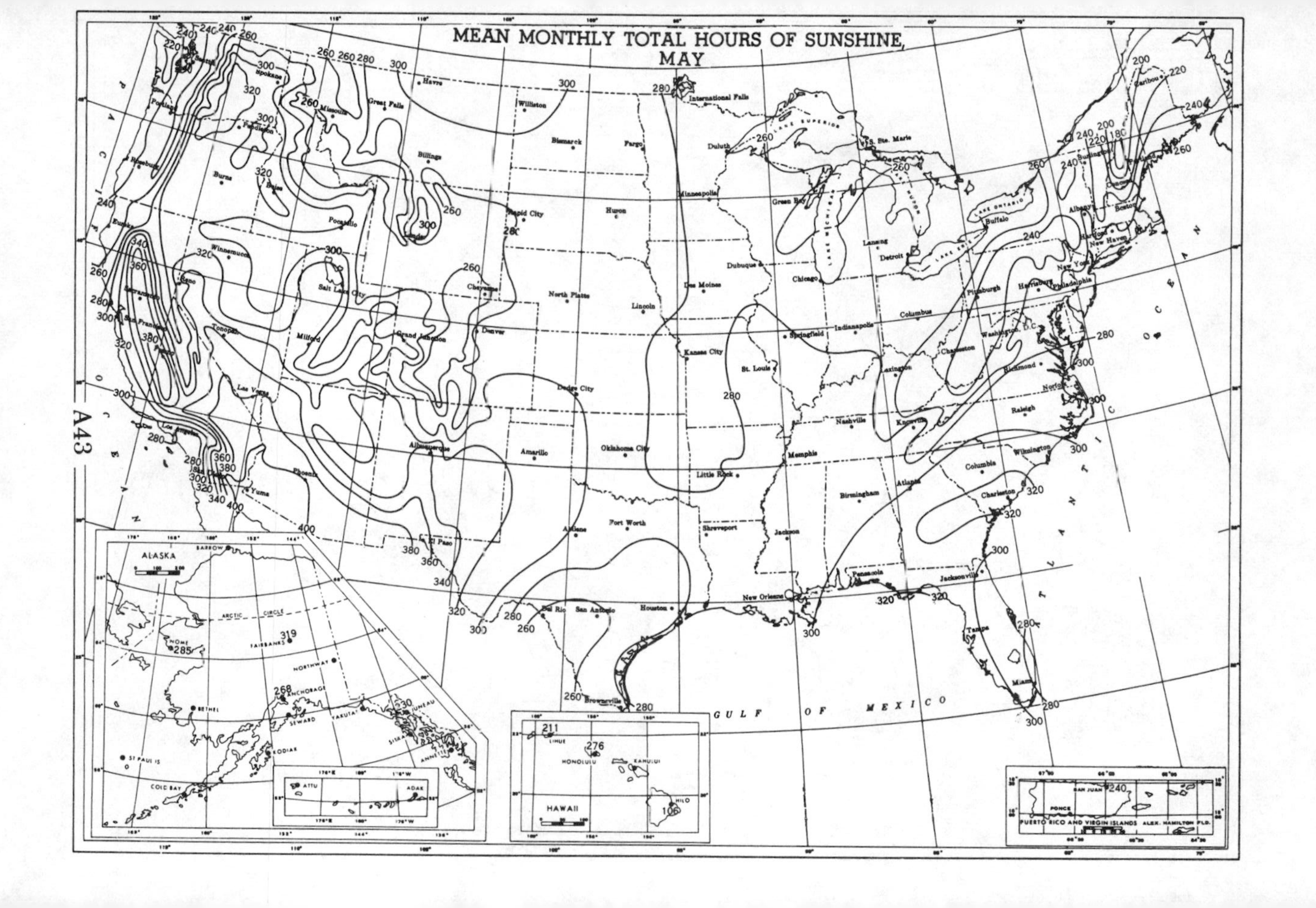
MEAN MONTHLY TOTAL HOURS OF SUNSHINE, MAY
GULF OF MEXICO
ALASKA
HAWAII
PUERTO RICO AND VIRGIN ISLANDS

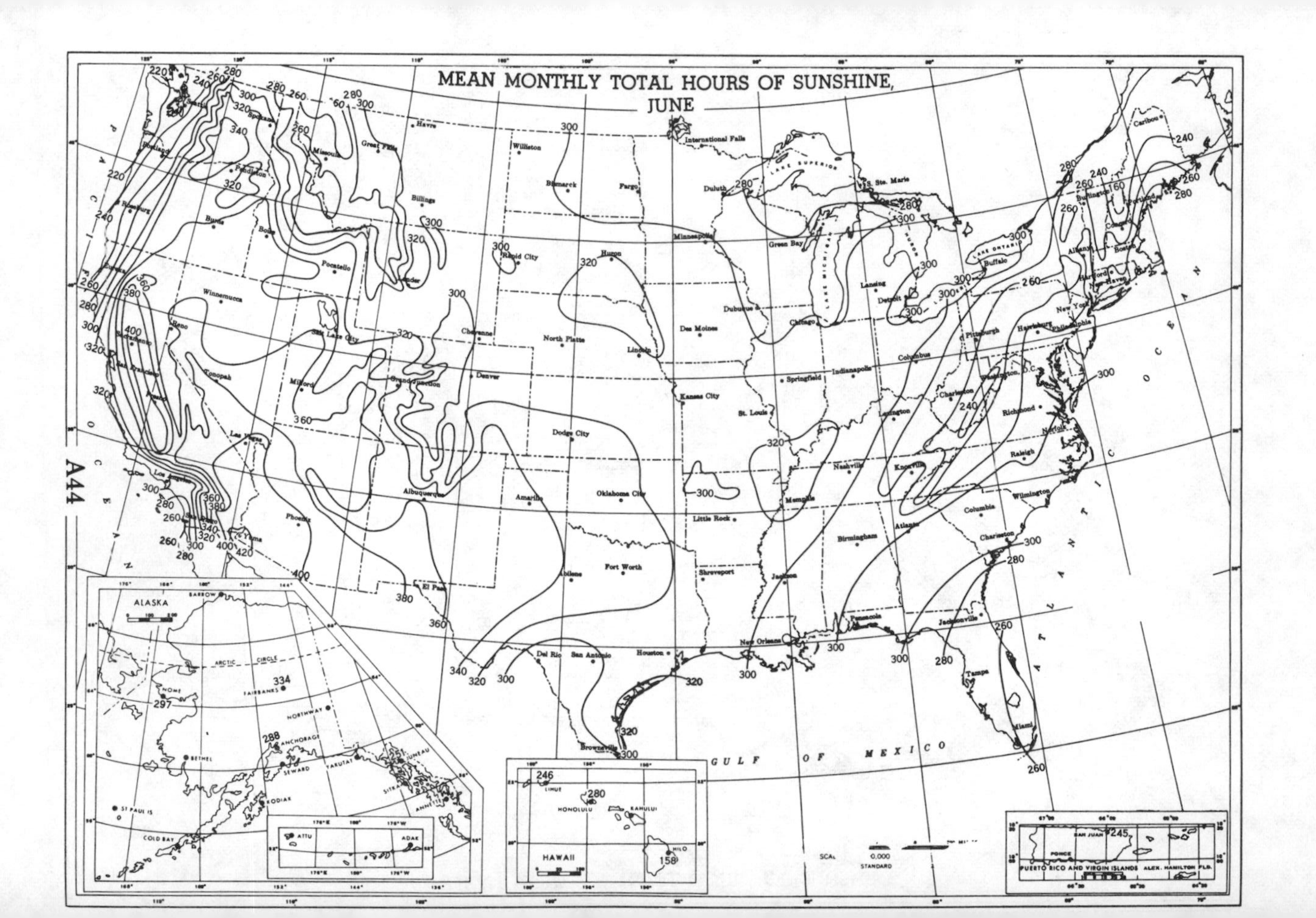
MEAN MONTHLY TOTAL HOURS OF SUNSHINE,
JUNE
PACIFIC OCEAN
ATLANTIC OCEAN
GULF OF MEXICO
ALASKA
ARCTIC CIRCLE
BARROW
NOME
297
FAIRBANKS
334
NORTHWAY
ANCHORAGE
288
BETHEL
SEWARD
YAKUTAT
JUNEAU
SITKA
ANNETTE
KODIAK
ST PAUL IS
COLD BAY
ATTU
ADAK
HAWAII
LIHUE
246
HONOLULU
280
KAHULUI
HILO
158
PUERTO RICO AND VIRGIN ISLANDS
SAN JUAN
245
PONCE
ALEX. HAMILTON FLD.
STANDARD
Seattle
Spokane
Portland
Pendleton
Roseburg
Eureka
Burns
Boise
Missoula
Great Falls
Havre
Billings
Pocatello
Lander
Winnemucca
Reno
Sacramento
San Francisco
Fresno
Tonopah
Salt Lake City
Milford
Las Vegas
Los Angeles
San Diego
Yuma
Phoenix
Grand Junction
Cheyenne
Denver
Albuquerque
El Paso
Williston
Bismarck
Fargo
Rapid City
Huron
North Platte
Lincoln
Dodge City
Amarillo
Oklahoma City
Abilene
Fort Worth
Del Rio
San Antonio
Houston
Brownsville
International Falls
Duluth
Minneapolis
Des Moines
Kansas City
Dubuque
Green Bay
Sault Ste. Marie
Chicago
Springfield
St. Louis
Little Rock
Shreveport
Memphis
Jackson
New Orleans
Lansing
Detroit
Indianapolis
Columbus
Lexington
Nashville
Knoxville
Birmingham
Atlanta
Pensacola
Jacksonville
Tampa
Miami
Buffalo
Pittsburgh
Charleston
Washington, D.C.
Richmond
Norfolk
Raleigh
Wilmington
Columbia
Harrisburg
Philadelphia
New York
Albany
Hartford
New Haven
Boston
Burlington
Concord
Portland
Caribou
LAKE SUPERIOR
LAKE MICHIGAN
HURON
LAKE ERIE
LAKE ONTARIO
220
240
260
280
300
320
340
360
380
400
420
160

MEAN MONTHLY TOTAL HOURS OF SUNSHINE, JULY

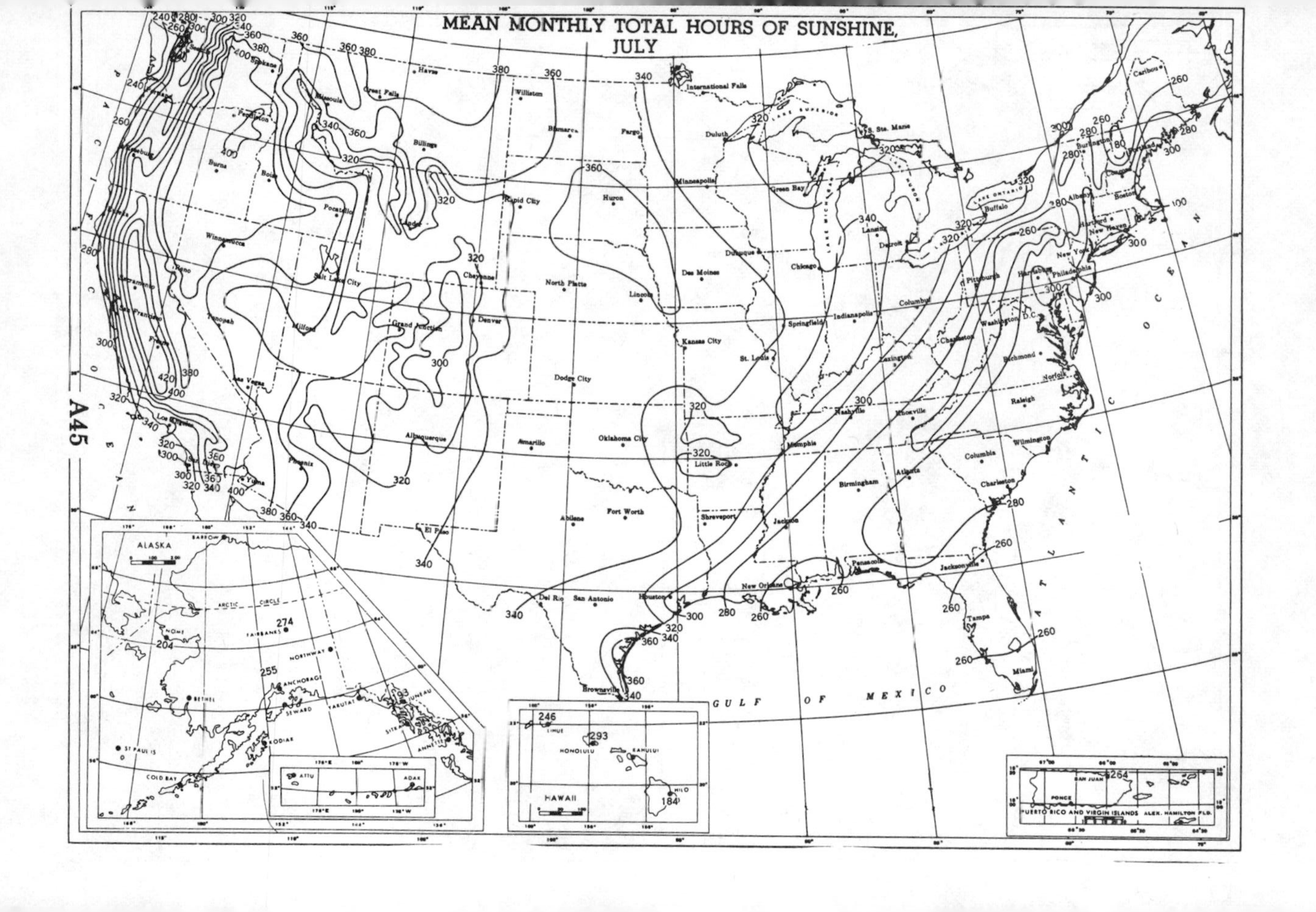

MEAN MONTHLY TOTAL HOURS OF SUNSHINE, AUGUST

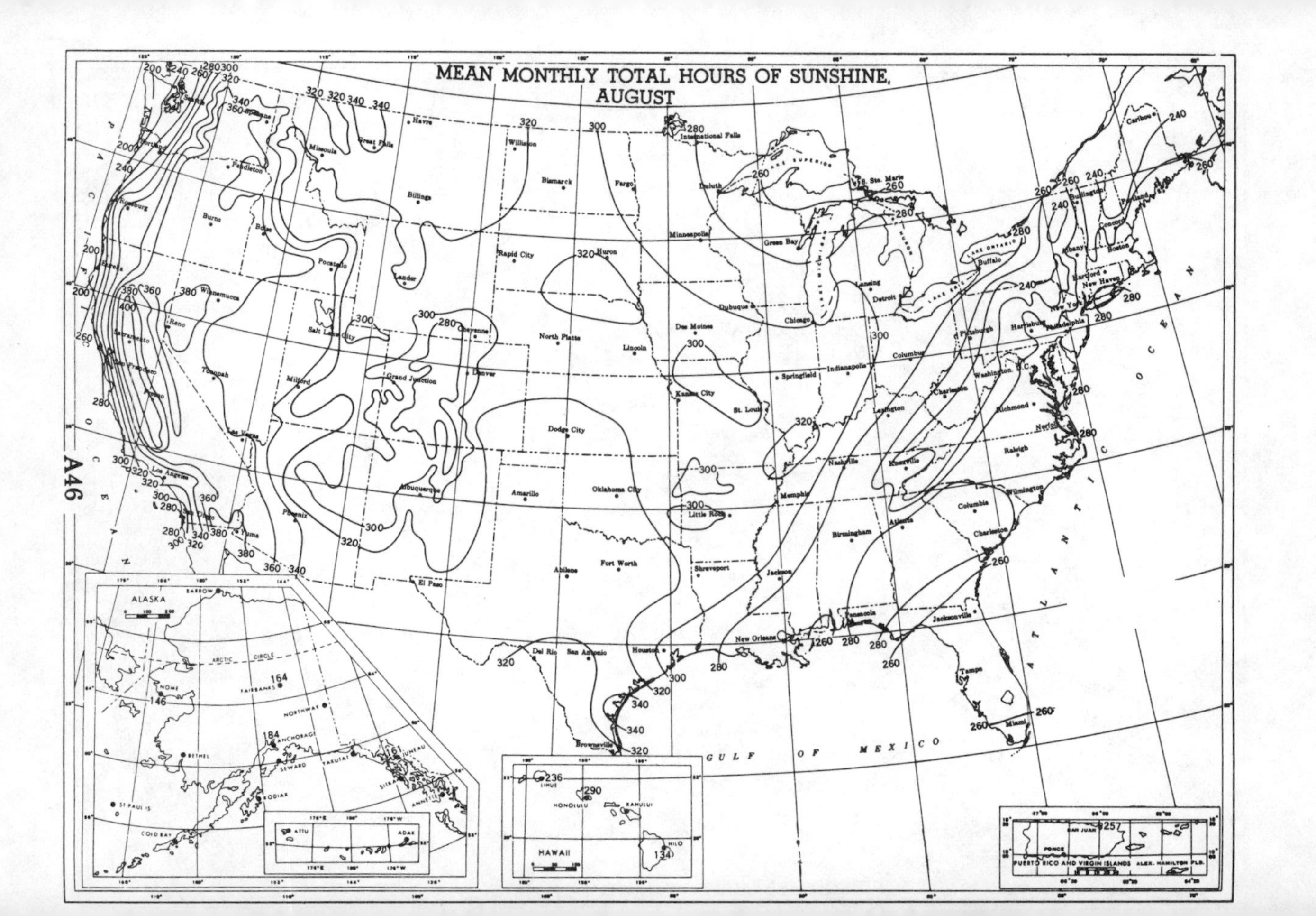

MEAN MONTHLY TOTAL HOURS OF SUNSHINE, SEPTEMBER

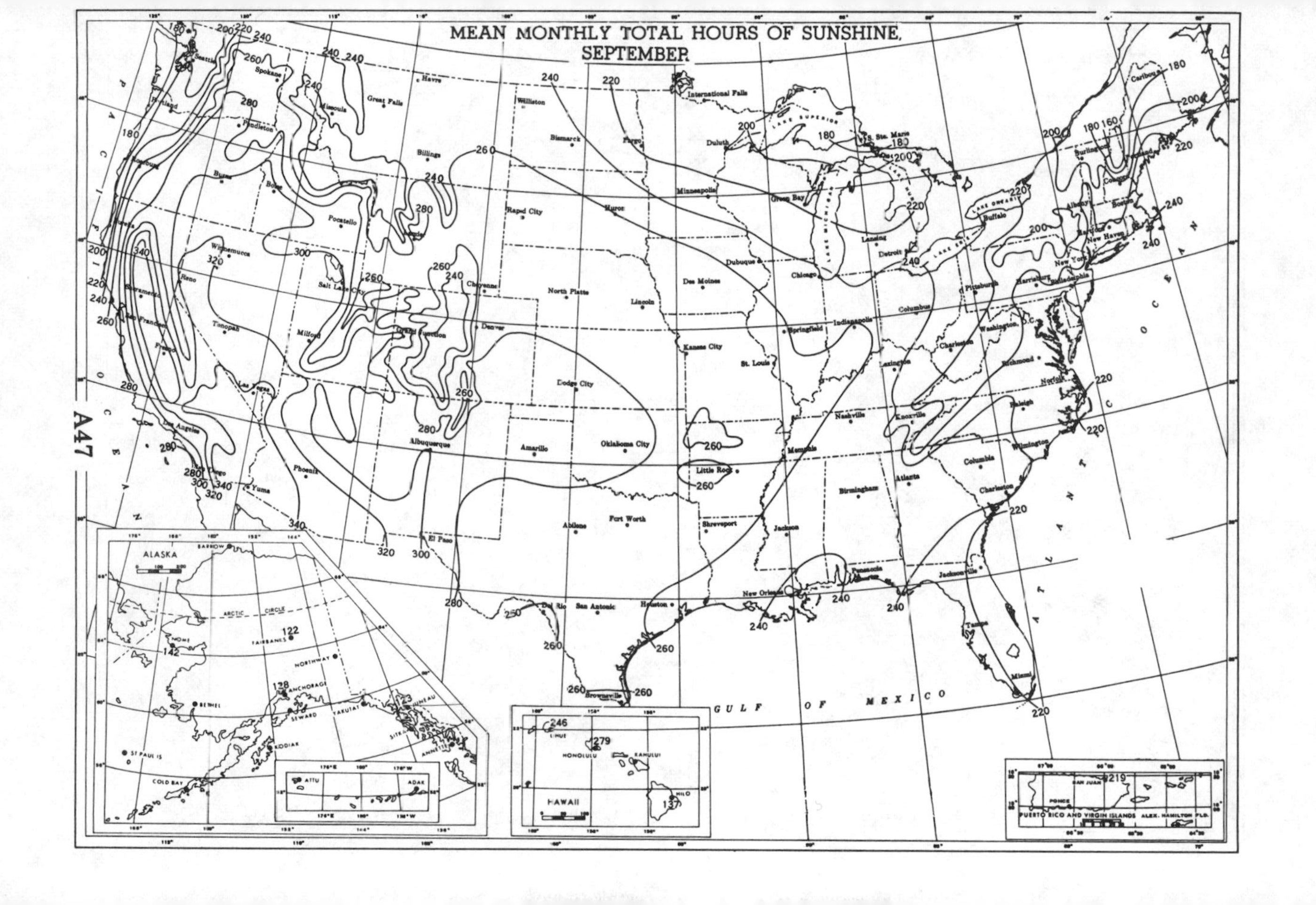

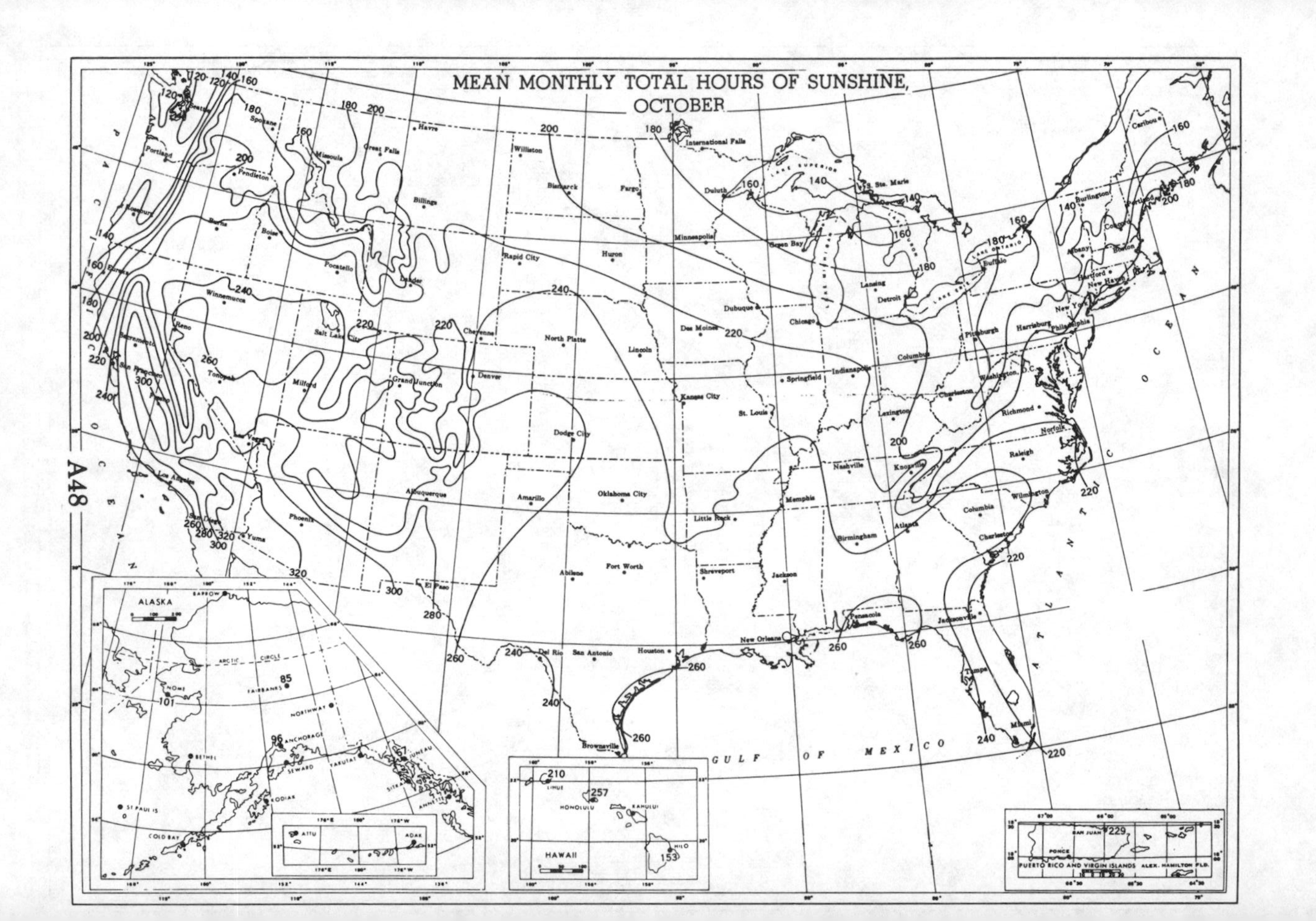
MEAN MONTHLY TOTAL HOURS OF SUNSHINE,
OCTOBER
PACIFIC OCEAN
ATLANTIC OCEAN
GULF OF MEXICO
ALASKA
ARCTIC CIRCLE
HAWAII
PUERTO RICO AND VIRGIN ISLANDS
Seattle
Portland
Roseburg
Eureka
Spokane
Pendleton
Burns
Boise
Missoula
Great Falls
Havre
Billings
Pocatello
Lander
Winnemucca
Reno
Sacramento
San Francisco
Fresno
Tonopah
Salt Lake City
Milford
Grand Junction
Cheyenne
Denver
Las Vegas
Los Angeles
San Diego
Yuma
Phoenix
Albuquerque
El Paso
Williston
Bismarck
Fargo
Rapid City
Huron
North Platte
Dodge City
Amarillo
Abilene
Fort Worth
Oklahoma City
Del Rio
San Antonio
Houston
Brownsville
International Falls
Duluth
Minneapolis
Green Bay
Sault Ste. Marie
Lansing
Detroit
Chicago
Dubuque
Des Moines
Lincoln
Kansas City
St. Louis
Springfield
Indianapolis
Columbus
Little Rock
Shreveport
Memphis
Jackson
Nashville
Birmingham
Atlanta
Knoxville
Lexington
Charleston
New Orleans
Pensacola
Jacksonville
Tampa
Miami
Columbia
Wilmington
Raleigh
Norfolk
Richmond
Washington, D.C.
Pittsburgh
Harrisburg
Philadelphia
New York
Buffalo
Albany
Hartford
New Haven
Boston
Concord
Burlington
Portland
Caribou
Barrow
Nome
Fairbanks
Northway
Anchorage
Bethel
Seward
Yakutat
Juneau
Sitka
Kodiak
Annette
St. Paul Is.
Cold Bay
Attu
Adak
Lihue
Honolulu
Kahului
Hilo
San Juan
Ponce
Alex. Hamilton Fld.

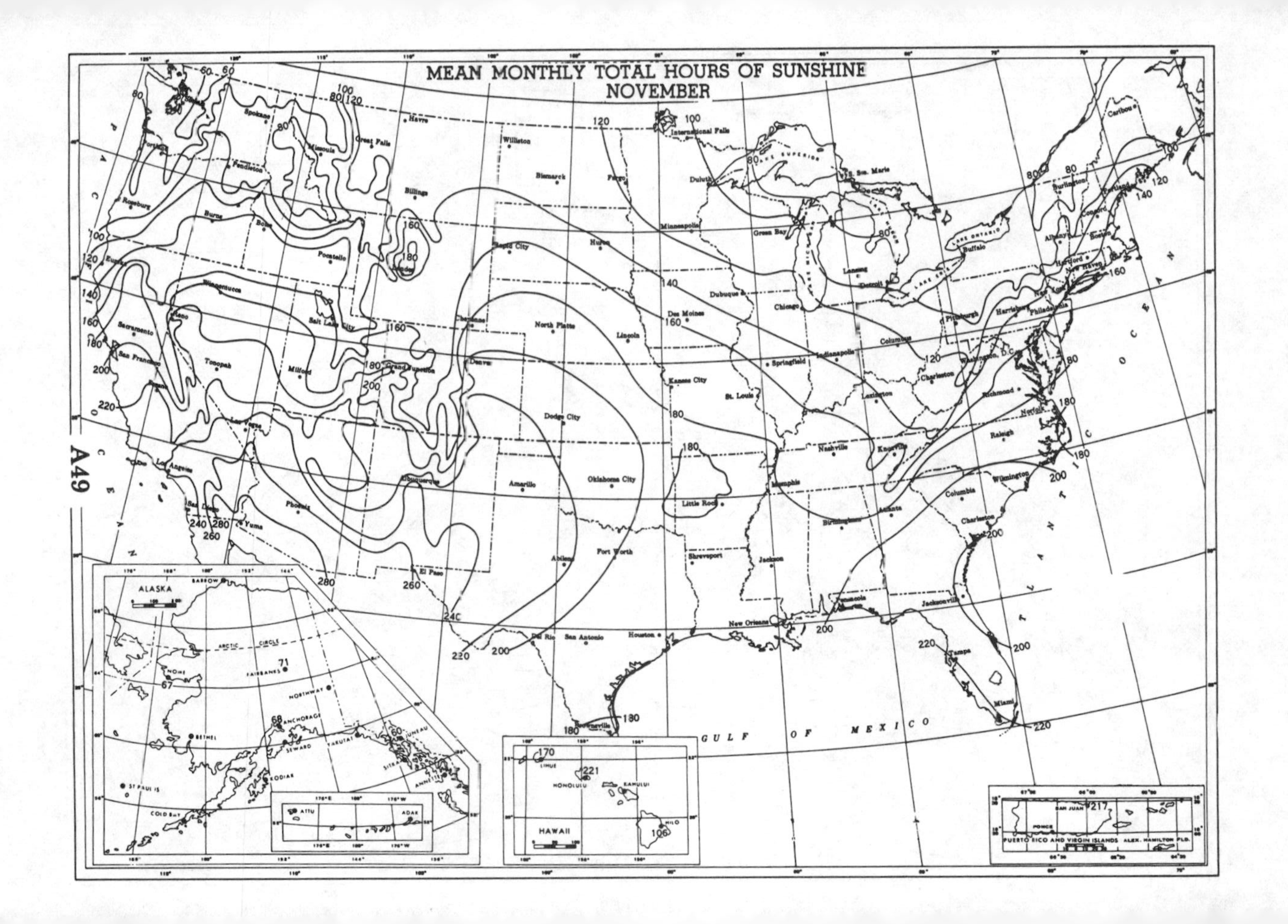
MEAN MONTHLY TOTAL HOURS OF SUNSHINE
NOVEMBER
ALASKA
HAWAII
PUERTO RICO AND VIRGIN ISLANDS
GULF OF MEXICO
ATLANTIC OCEAN
PACIFIC OCEAN

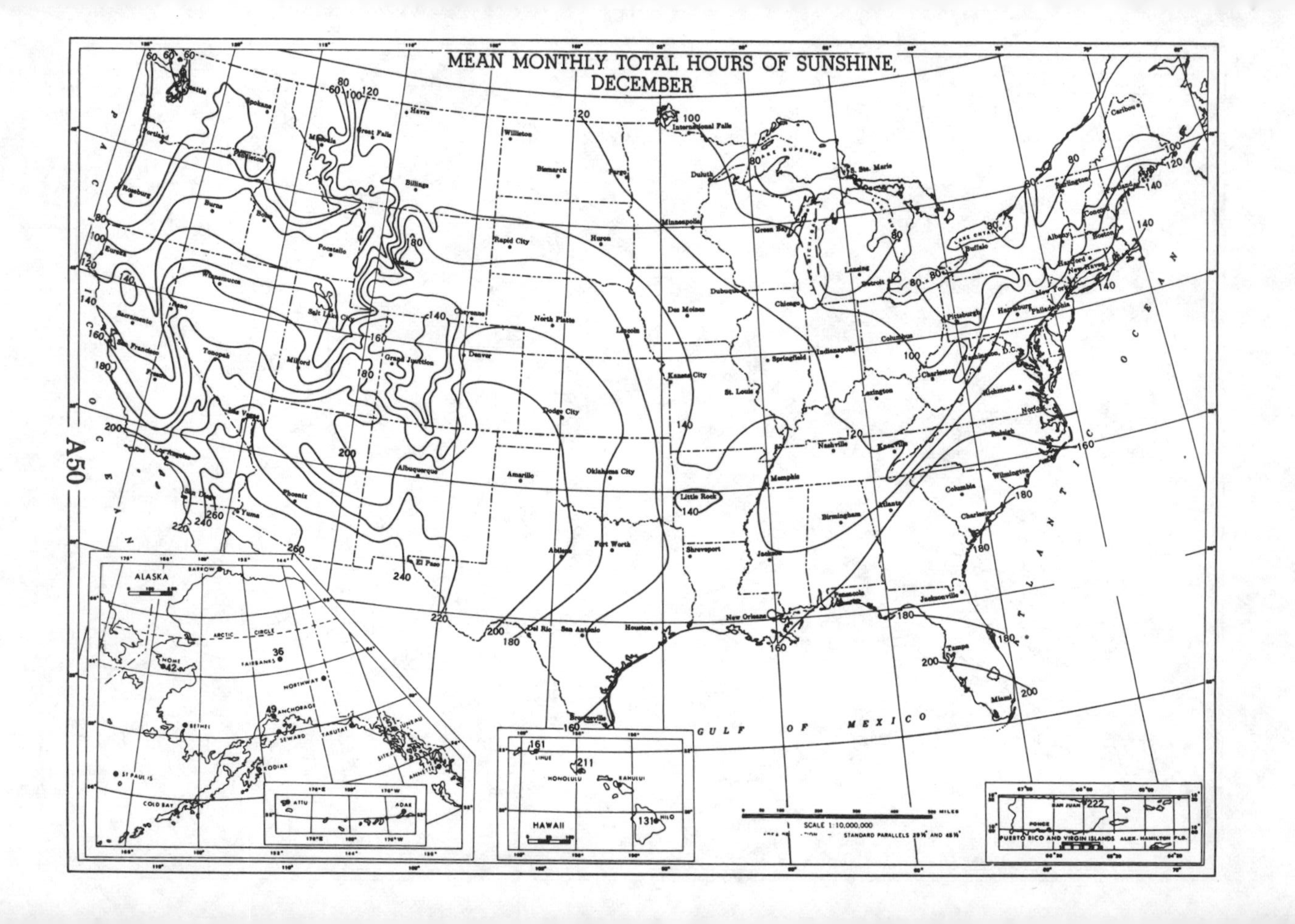
MEAN MONTHLY TOTAL HOURS OF SUNSHINE,
DECEMBER
GULF OF MEXICO
ALASKA
HAWAII
PUERTO RICO AND VIRGIN ISLANDS
SCALE 1:10,000,000

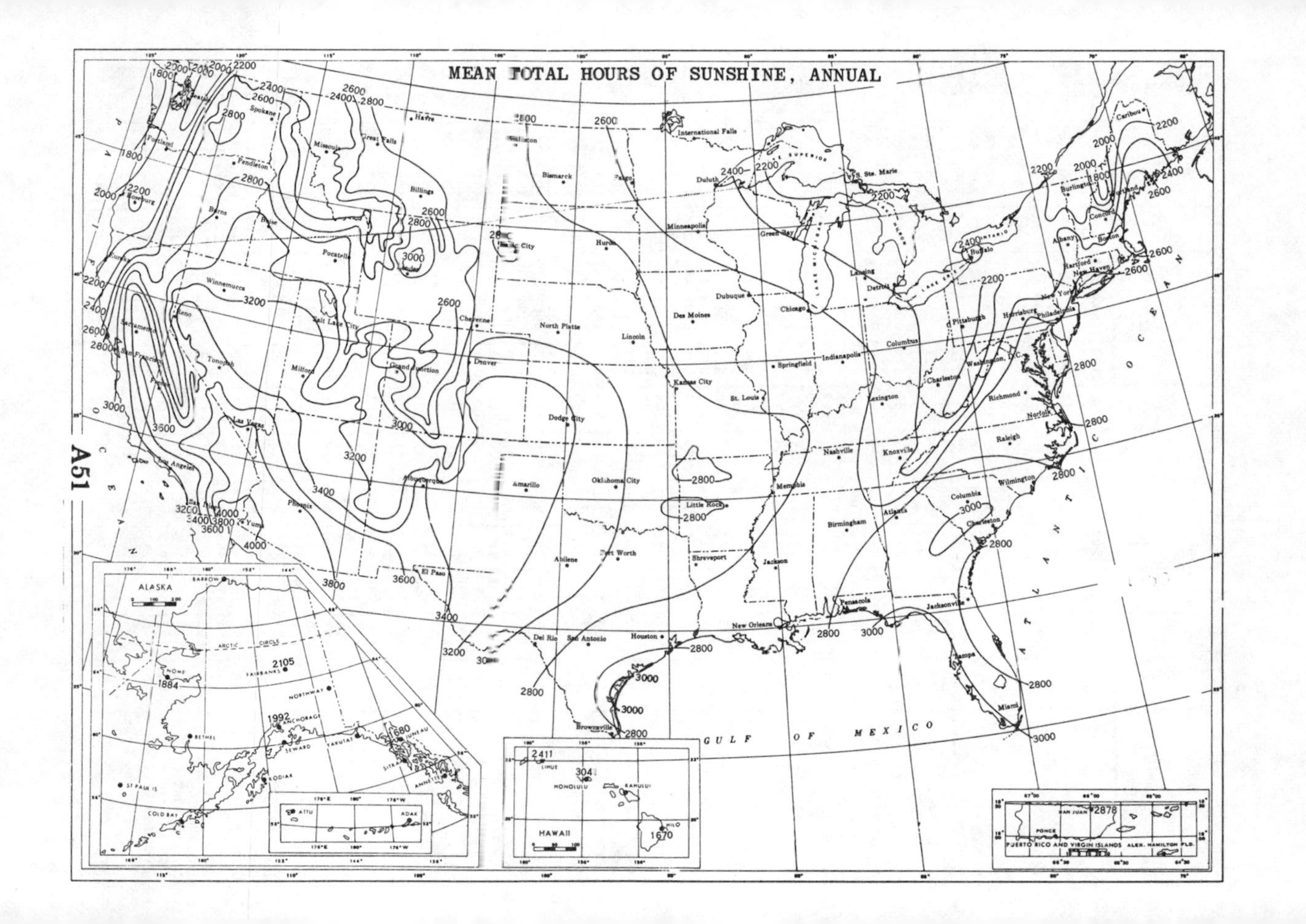
MEAN TOTAL HOURS OF SUNSHINE, ANNUAL
ALASKA
ARCTIC CIRCLE
2105
1884
1992
1680
HAWAII
2411
3041
1670
PUERTO RICO AND VIRGIN ISLANDS
2878
GULF OF MEXICO
ATLANTIC OCEAN
PACIFIC OCEAN

MEAN TOTAL NUMBER OF HOURS OF SUNSHINE

STATE AND STATION	YEARS	JAN.	FEB.	MAR.	APR.	MAY	JUNE	JULY	AUG.	SEPT.	OCT.	NOV.	DEC.	ANNUAL
ALA. BIRMINGHAM	30	138	152	207	248	293	294	269	265	244	234	182	136	2662
MOBILE	22	157	158	212	253	301	289	249	259	235	254	195	146	2708
MONTGOMERY	30	160	168	227	267	317	311	288	290	260	250	200	156	2894
ALASKA ANCHORAGE	19	78	114	210	254	268	288	255	184	128	96	68	49	1992
FAIRBANKS	20	54	120	224	302	319	334	274	164	122	85	71	36	2105
JUNEAU	29	71	102	171	200	230	251	193	161	123	67	60	51	1680
NOME	27	72	109	193	226	285	297	204	146	142	101	67	42	1884
ARIZ. PHOENIX	30	248	244	314	346	404	404	377	351	334	307	267	236	3832
PRESCOTT	14	222	230	293	323	378	392	323	305	315	286	254	228	3549
TUCSON	13	255	266	317	350	399	394	329	329	335	317	280	258	3829
YUMA	30	258	266	337	365	419	420	404	380	351	330	285	262	4077
ARK. FT. SMITH	30	146	156	202	234	268	303	321	305	261	230	174	147	2747
LITTLE ROCK	30	143	158	213	243	291	316	321	316	265	251	181	142	2840
CALIF. EUREKA	30	120	138	180	209	247	261	244	205	195	164	127	108	2198
FRESNO	29	153	192	283	330	389	418	435	406	355	306	221	144	3632
LOS ANGELES	30	224	217	273	264	292	299	352	336	295	263	249	220	3284
RED BLUFF	15	156	186	246	302	366	396	438	407	341	277	199	154	3468
SACRAMENTO	30	134	169	255	300	367	405	437	406	347	283	197	122	3422
SAN DIEGO	30	216	212	262	242	261	253	293	277	255	234	236	217	2958
SAN FRANCISCO	30	165	182	251	281	314	330	300	272	267	243	198	156	2959
COLO. DENVER	30	207	205	247	252	281	311	321	297	274	246	200	192	3033
GRAND JUNCTION	30	169	182	243	265	314	350	349	311	291	255	198	168	3095
PUEBLO	30	224	217	261	271	299	340	349	318	290	265	225	211	3270
CONN. HARTFORD	30	141	166	206	223	267	285	299	268	220	193	137	136	2541
NEW HAVEN	30	155	178	215	234	274	291	309	284	238	215	157	154	2704
D. C. WASHINGTON	30	138	160	205	226	267	288	291	264	233	207	162	135	2576
FLA. APALACHICOLA	26	193	195	233	274	328	296	273	259	236	263	216	175	2941
JACKSONVILLE	30	192	189	241	267	296	260	255	248	199	205	191	170	2713
KEY WEST	30	229	238	285	296	307	273	277	269	236	237	226	225	3098
LAKELAND	7	204	186	222	251	285	268	252	242	203	209	212	198	2732
MIAMI	30	222	227	266	275	280	251	267	263	216	215	212	209	2903
PENSACOLA	30	175	180	232	270	311	302	278	284	249	265	206	166	2918
TAMPA	30	223	220	260	283	320	275	257	252	232	243	227	209	3001
GA. ATLANTA	25	154	165	218	266	309	304	284	285	247	241	188	160	2821
MACON	30	177	178	235	279	321	314	292	295	253	236	202	168	2950
SAVANNAH	30	175	173	229	274	307	279	267	256	212	216	197	167	2752
HAWAII HILO	7	153	135	161	112	106	158	184	134	137	153	106	131	1670
HONOLULU	30	227	202	250	255	276	280	293	290	279	257	221	211	3041
LIHUE	10	171	162	176	176	211	246	246	236	246	210	170	161	2411
IDAHO BOISE	30	116	144	218	274	322	352	412	378	311	232	143	104	3006
POCATELLO	30	111	143	211	255	300	338	380	347	296	230	145	108	2864
ILL. CAIRO	15	124	160	218	254	298	324	345	336	279	254	181	145	2918
CHICAGO	30	126	142	199	221	274	300	333	299	247	216	136	118	2611
MOLINE	18	132	139	189	214	255	279	337	300	251	214	130	123	2563
PEORIA	30	134	149	198	229	273	303	336	299	259	222	149	122	2673
SPRINGFIELD	30	127	149	193	224	282	304	346	312	266	225	152	122	2702
IND. EVANSVILLE	30	123	145	199	237	294	322	342	318	274	236	156	120	2766
FT. WAYNE	30	113	136	191	217	281	310	342	306	242	210	120	102	2570
INDIANAPOLIS	30	118	140	193	227	278	313	342	313	265	222	139	118	2668
TERRE HAUTE	24	125	148	189	231	274	302	341	305	253	235	150	122	2675
IOWA BURLINGTON	19	148	165	217	241	284	315	353	327	270	243	175	147	2885
CHARLES CITY	22	137	157	190	226	258	285	336	290	241	207	130	115	2572
DES MOINES	30	155	170	203	236	276	303	346	299	263	227	156	136	2770
SIOUX CITY	30	164	177	216	254	300	320	363	320	270	236	160	146	2926
KAN. CONCORDIA	30	180	172	214	243	281	315	348	308	249	245	189	172	2916
DODGE CITY	30	205	191	249	268	305	335	359	335	290	266	218	198	3219
TOPEKA	18	159	160	193	215	260	287	310	304	263	229	173	149	2702
WICHITA	30	187	186	233	254	291	321	350	325	277	245	206	182	3057
KY. LOUISVILLE	30	115	135	188	221	283	303	324	295	256	219	148	114	2601
LA. NEW ORLEANS	30	160	158	213	247	292	287	260	269	241	260	200	157	2744
SHREVEPORT	19	151	172	214	240	298	332	339	322	289	273	208	177	3015
MAINE EASTPORT	22	133	151	196	201	245	248	275	260	205	175	105	115	2309
PORTLAND	30	155	174	213	226	268	286	312	294	229	202	146	148	2653
MD. BALTIMORE	30	148	170	211	229	270	295	299	272	238	212	164	145	2653
MASS. BLUE HILL OBS.	10	125	136	165	182	233	248	266	241	211	181	134	135	2257
BOSTON	30	148	168	212	222	263	283	300	280	232	207	152	148	2615
NANTUCKET	22	128	156	214	227	278	284	291	279	242	208	149	129	2585

MEAN TOTAL NUMBER OF HOURS OF SUNSHINE

STATE AND STATION	YEARS	JAN.	FEB.	MAR.	APR.	MAY	JUNE	JULY	AUG.	SEPT.	OCT.	NOV.	DEC.	ANNUAL
MICH. ALPENA	24	86	124	198	228	261	303	339	285	204	159	70	67	2324
DETROIT	30	90	128	180	212	263	295	321	284	226	189	98	89	2375
LANSING	30	84	119	175	215	272	305	344	294	228	182	87	73	2378
ESCANABA	30	112	148	204	226	266	283	316	267	198	162	90	94	2366
GRAND RAPIDS	30	74	117	178	218	277	308	349	304	231	188	92	70	2406
MARQUETTE	30	78	113	172	207	248	268	305	251	186	142	68	66	2104
SAULT STE. MARIE	30	83	123	187	217	252	269	309	256	165	133	61	62	2117
MINN. DULUTH	30	125	163	221	235	268	282	328	277	203	166	100	107	2475
MINNEAPOLIS	30	140	166	200	231	272	302	343	296	237	193	115	112	2607
MISS. JACKSON	12	130	147	199	244	280	287	279	287	235	223	185	150	2646
VICKSBURG	30	136	141	199	232	284	304	291	297	254	244	183	140	2705
MO. COLUMBIA	30	147	164	207	232	281	296	341	298	262	225	166	138	2757
KANSAS CITY	30	154	170	211	235	278	313	347	308	266	235	178	151	2846
ST. JOSEPH	23	154	165	211	231	274	301	347	287	260	224	168	144	2766
ST. LOUIS	30	137	152	202	235	283	301	325	289	256	223	166	125	2694
SPRINGFIELD	30	145	164	213	238	278	305	342	310	269	233	183	140	2820
MONT. BILLINGS	21	140	154	208	236	283	301	372	332	258	213	136	129	2762
GREAT FALLS	19	154	176	245	261	299	299	381	342	256	206	132	133	2884
HAVRE	30	136	174	234	268	311	312	384	339	260	202	132	122	2874
HELENA	30	138	168	215	241	292	292	342	336	258	202	137	121	2742
MISSOULA	25	85	109	167	209	261	260	378	328	246	178	90	66	2377
NEBR. LINCOLN	30	173	172	213	244	287	316	356	309	266	237	174	160	2907
NORTH PLATTE	30	181	179	221	246	282	310	343	304	264	242	184	169	2925
OMAHA	30	172	188	222	259	305	332	379	311	270	248	166	145	2997
VALENTINE	30	185	194	229	252	296	323	369	326	275	242	174	172	3037
NEV. ELY	22	186	197	262	260	300	354	359	344	303	255	204	187	3211
LAS VEGAS	8	239	251	314	336	386	411	383	364	345	301	258	250	3838
RENO	30	185	199	267	306	354	376	414	391	336	273	212	170	3483
WINNEMUCCA	30	142	155	207	255	312	346	395	375	316	242	177	139	3061
N. H. CONCORD	23	136	153	192	196	229	261	286	260	214	179	122	126	2354
MT. WASHINGTON OBS.	18	94	98	133	141	162	145	150	143	139	159	89	87	1540
N. J. ATLANTIC CITY	30	151	173	210	233	273	287	298	271	239	218	177	153	2683
TRENTON	30	145	168	203	235	277	294	309	273	239	208	160	142	2653
N. MEX. ALBUQUERQUE	30	221	218	273	299	343	365	340	317	299	279	245	219	3418
ROSWELL	21	218	223	286	306	330	333	341	313	266	266	242	216	3340
N. Y. ALBANY	30	125	151	194	213	266	301	317	286	224	192	115	112	2496
BINGHAMTON	30	94	119	151	170	226	256	266	230	184	158	92	79	2025
BUFFALO	30	110	125	180	212	274	319	338	297	239	183	97	84	2458
NEW YORK	30	154	171	213	237	268	289	302	271	235	213	169	155	2677
ROCHESTER	30	93	123	172	209	274	314	333	294	224	173	97	86	2392
SYRACUSE	30	[illegible]	[illegible]	[illegible]	[illegible]	[illegible]	[illegible]	[illegible]	[illegible]	[illegible]	[illegible]	81	74	[illegible]
N. C. ASHEVILLE	30	146	161	211	247	289	292	268	250	235	222	179	140	[illegible]
CAPE HATTERAS	9	152	168	206	259	293	301	286	265	214	202	169	154	2669
CHARLOTTE	30	165	177	230	267	313	316	291	277	247	243	198	167	2891
GREENSBORO	30	157	171	217	231	298	302	287	272	243	236	190	163	2767
RALEIGH	29	154	168	220	255	290	284	277	253	224	215	184	156	2680
WILMINGTON	30	179	180	237	279	314	312	286	273	237	238	206	178	2919
N. DAK. BISMARCK	30	141	170	205	236	279	294	358	307	243	198	130	125	2686
DEVILS LAKE	30	150	177	220	250	291	297	352	302	230	198	123	124	2714
FARGO	30	132	170	210	232	283	288	343	293	222	187	112	114	2586
WILLISTON	29	141	168	215	260	305	312	377	328	247	206	131	129	2819
OHIO CINCINNATI (ABBE)	30	115	137	186	222	273	309	323	295	253	205	138	118	2574
CLEVELAND	30	79	111	167	209	274	301	325	288	235	187	99	77	2352
COLUMBUS	30	112	132	177	215	270	296	323	291	250	210	131	101	2508
DAYTON	10	114	136	195	222	281	313	323	307	268	229	152	124	2664
SANDUSKY	30	100	128	183	229	285	312	343	302	248	201	111	91	2533
TOLEDO	30	93	120	170	203	263	296	331	298	241	196	106	92	2409
OKLA. OKLAHOMA CITY	29	175	182	235	253	290	329	352	331	282	243	201	175	3048
TULSA	18	152	164	200	213	244	287	314	308	281	241	207	172	2783
OREG. BAKER	22	118	143	198	251	302	313	406	368	289	215	132	100	2835
PORTLAND	30	77	97	142	203	246	249	329	275	218	134	87	65	2122
ROSEBURG	30	69	96	148	205	257	278	369	329	255	146	81	50	2283
PA. HARRISBURG	30	132	160	203	230	277	297	319	282	233	200	140	131	2604
PHILADELPHIA	30	142	166	203	231	270	281	288	253	225	205	158	142	2564
PITTSBURGH	25	89	114	163	200	239	260	283	250	234	180	114	76	2202
READING	30	133	151	195	220	259	275	293	259	219	198	144	127	2473
SCRANTON	30	108	138	178	199	251	269	290	249	213	183	120	105	2303
R. I. PROVIDENCE	30	145	168	211	221	271	285	292	267	226	207	153	143	2589

MEAN TOTAL NUMBER OF HOURS OF SUNSHINE

STATE AND STATION	YEARS	JAN.	FEB.	MAR.	APR.	MAY	JUNE	JULY	AUG.	SEPT.	OCT.	NOV.	DEC.	ANNUAL
S. C. CHARLESTON	30	188	189	243	284	323	308	297	281	244	239	210	187	2993
COLUMBIA	30	173	183	233	274	312	312	291	283	243	242	202	166	2914
GREENVILLE	26	166	176	227	274	307	300	278	274	239	232	192	157	2822
S. DAK. HURON	30	153	177	213	250	295	321	367	320	260	212	142	134	2844
RAPID CITY	30	164	182	222	245	278	300	348	317	266	228	164	144	2858
TENN. CHATTANOOGA	30	126	146	187	239	290	295	278	266	247	220	169	128	2591
KNOXVILLE	30	124	144	189	237	281	288	277	248	237	213	157	120	2515
MEMPHIS	30	135	152	204	244	296	321	319	314	261	243	180	139	2808
NASHVILLE	30	123	142	196	241	285	308	292	279	250	224	168	126	2634
TEX. ABILENE	13	190	199	250	259	290	347	335	322	276	245	223	201	3137
AMARILLO	30	207	199	258	276	305	338	350	328	288	260	229	205	3243
AUSTIN	30	148	152	207	221	266	302	331	320	261	242	180	160	2790
BROWNSVILLE	30	147	152	187	210	272	297	326	311	246	252	165	151	2716
CORPUS CHRISTI	24	160	165	212	237	295	329	366	341	276	264	194	164	3003
DALLAS	30	155	159	220	238	279	326	341	325	274	240	191	163	2911
DEL RIO	27	173	173	230	237	259	279	331	319	252	240	195	178	2866
EL PASO	30	234	236	299	329	373	369	336	327	300	287	257	236	3583
GALVESTON	30	151	149	203	230	288	322	305	292	257	264	199	151	2811
HOUSTON	30	144	141	193	212	266	298	294	281	238	239	181	146	2633
PORT ARTHUR	30	153	149	209	235	292	317	285	281	252	256	191	148	2768
SAN ANTONIO	30	148	153	213	224	258	292	325	307	261	241	183	160	2765
UTAH SALT LAKE CITY	30	137	155	227	269	329	358	377	346	306	249	171	135	3059
VT. BURLINGTON	30	103	127	184	185	244	270	291	266	199	152	77	80	2178
VA. LYNCHBURG	26	153	169	216	243	288	297	288	264	235	217	177	158	2705
NORFOLK	30	156	174	223	257	304	311	296	282	237	220	182	161	2803
RICHMOND	30	144	166	211	248	280	296	286	263	230	211	176	152	2663
WASH. NORTH HEAD	22	76	97	135	182	221	214	226	186	170	123	87	66	1783
SEATTLE	30	74	99	154	201	247	234	304	248	197	122	77	62	2019
SPOKANE	30	78	120	197	262	308	309	397	350	264	177	86	57	2605
TATOOSH ISLAND	30	70	100	135	182	229	217	235	190	175	129	71	60	1793
WALLA WALLA	30	72	106	194	262	317	335	411	367	280	198	92	51	2685
W. VA. ELKINS	24	110	119	158	198	227	256	225	236	211	186	131	103	2160
PARKERSBURG	30	91	111	155	200	252	277	286	264	230	189	117	93	2265
WIS. GREEN BAY	30	121	148	194	210	251	279	314	266	213	176	110	106	2388
MADISON	30	126	147	196	214	258	285	336	288	230	198	116	108	2502
MILWAUKEE	30	116	134	191	218	267	293	340	292	235	193	125	106	2510
WYO. CHEYENNE	30	191	197	243	237	259	304	318	286	265	242	188	170	2900
LANDER	30	200	208	260	264	301	340	361	326	280	233	186	185	3144
SHERIDAN	30	160	179	226	245	286	303	367	333	266	221	153	145	2884
P. R. SAN JUAN	30	231	229	273	252	240	245	264	257	219	229	217	222	2878

The smoothed isolines on these charts and the data in this tabulation are based on Weather Bureau records from black-bulb type sunshine recorders. These values are those made during the 1931-60 period.

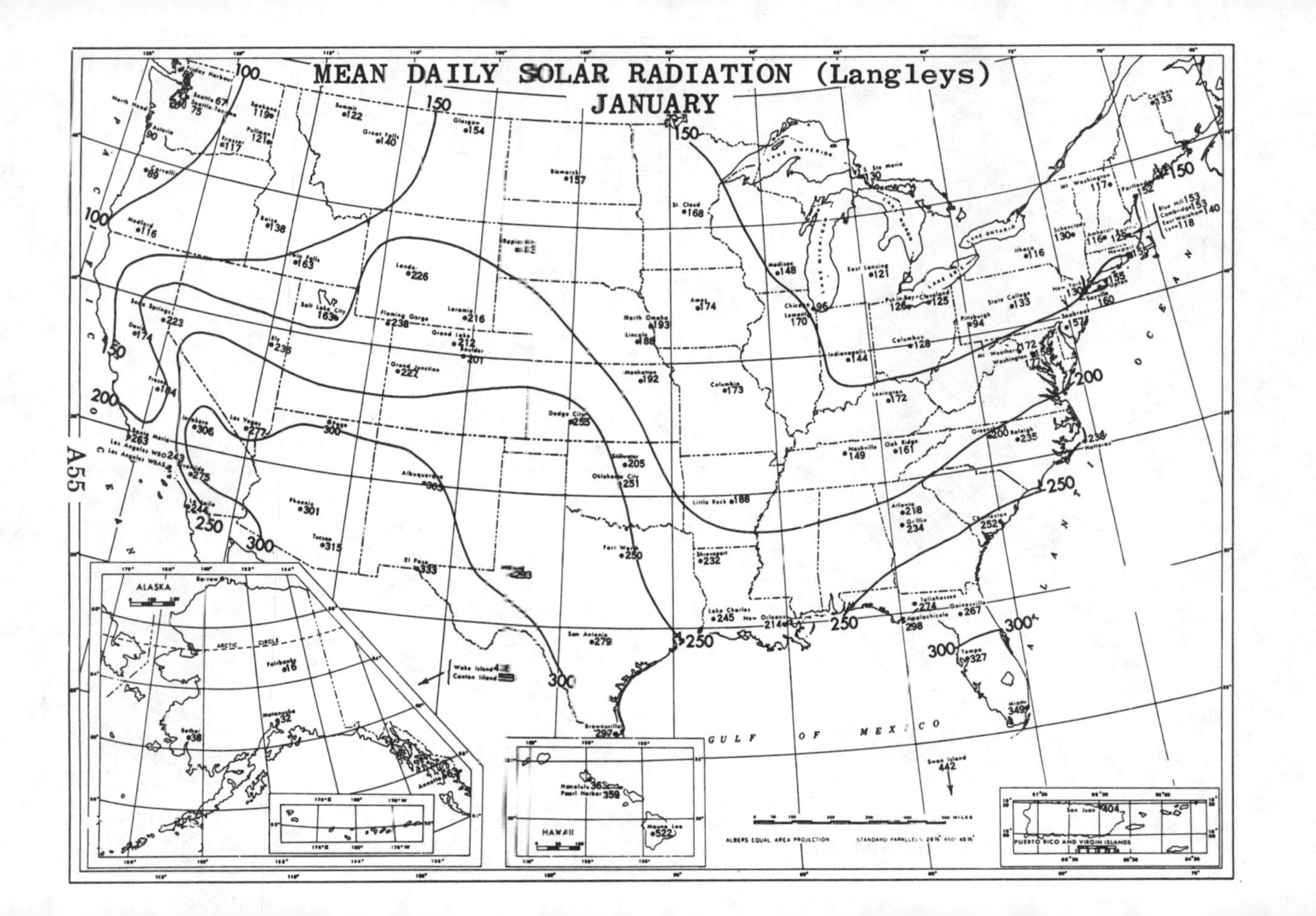
MEAN DAILY SOLAR RADIATION (Langleys)
JANUARY
GULF OF MEXICO
ALASKA
HAWAII
PUERTO RICO AND VIRGIN ISLANDS
ALBERS EQUAL AREA PROJECTION

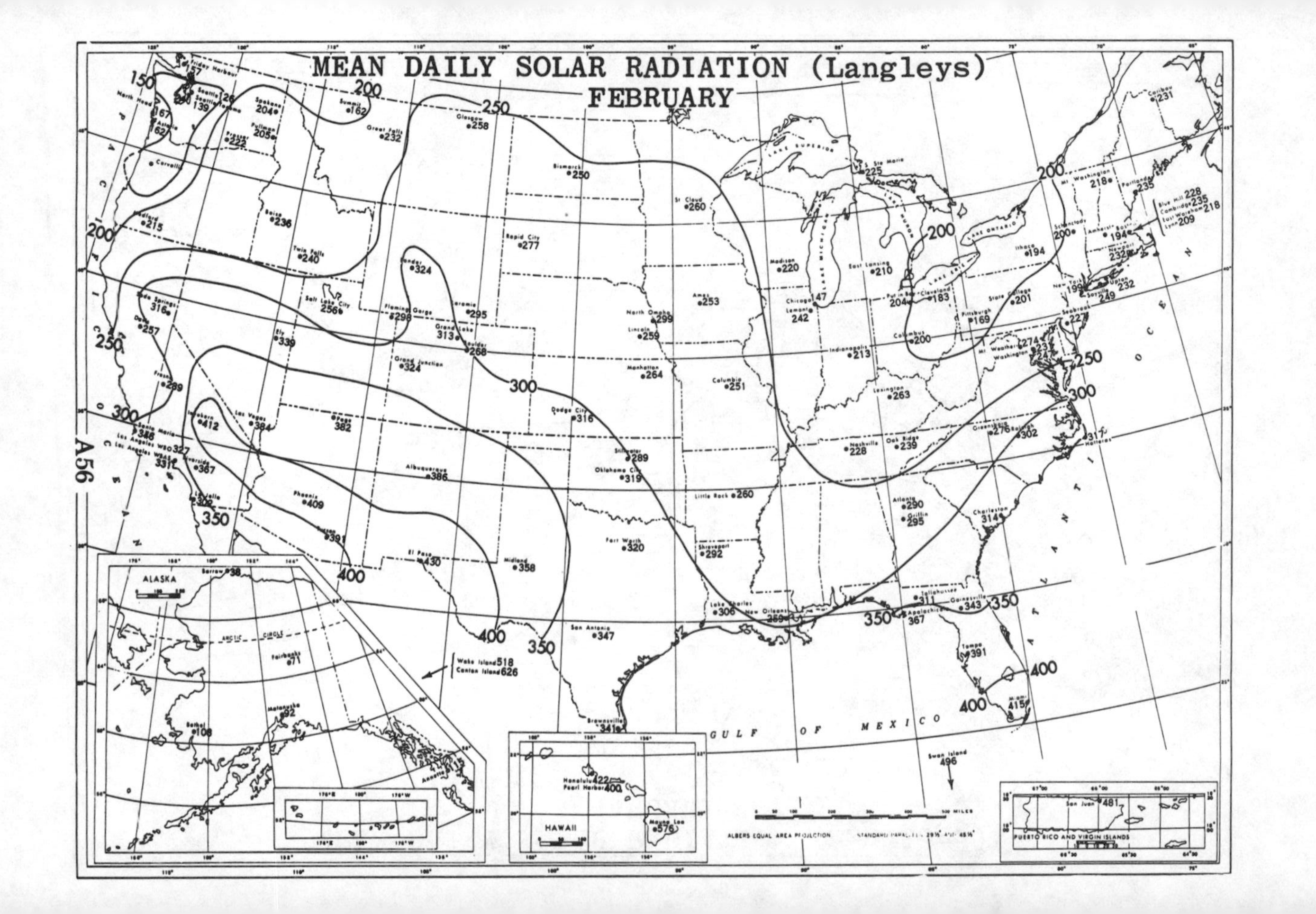
MEAN DAILY SOLAR RADIATION (Langleys)
FEBRUARY
GULF OF MEXICO
ATLANTIC OCEAN
PACIFIC OCEAN
ALASKA
HAWAII
PUERTO RICO AND VIRGIN ISLANDS
ALBERS EQUAL AREA PROJECTION
LAKE SUPERIOR
LAKE MICHIGAN
LAKE ONTARIO
LAKE ERIE
Wake Island 518
Canton Island 626
Swan Island 496
Honolulu 422
Pearl Harbor 400
Mauna Loa 576
San Juan 481
Barrow 38
Fairbanks 71
Matanuska 92
Bethel 108
Caribou 231
Portland 235
Mt Washington 218
Blue Hill 228
Cambridge 235
East Wareham 218
Lynn 209
Boston 194
Schenectady 200
Ithaca 194
Newport 232
New York 199
Upton 232
Sayville 249
Seabrook 227
State College 201
Pittsburgh 169
Cleveland 183
Put in Bay 204
Columbus 200
East Lansing 210
Sault Ste Marie 225
Madison 220
Chicago 147
Lemont 242
Indianapolis 213
Lexington 263
Nashville 228
Oak Ridge 239
Greensboro 276
Raleigh 302
Hatteras 317
Washington 274
Charleston 314
Atlanta 290
Griffin 295
Tallahassee 311
Apalachicola 367
Gainesville 343
Tampa 391
Miami 415
New Orleans 269
Lake Charles 306
Little Rock 260
Shreveport 292
Brownsville 341
San Antonio 347
Fort Worth 320
Oklahoma City 319
Stillwater 289
Dodge City 316
Manhattan 264
Columbia 251
Ames 253
North Omaha 299
Lincoln 259
St Cloud 260
Bismarck 250
Rapid City 277
Glasgow 258
Great Falls 232
Summit 162
Spokane 204
Pullman 205
Prosser 222
Seattle 126
Seattle Tacoma 139
North Head 167
Astoria 162
Friday Harbour 157
Corvallis
Medford 215
Boise 236
Twin Falls 240
Salt Lake City 256
Lander 324
Laramie 295
Flaming Gorge 298
Grand Lake 313
Boulder 268
Grand Junction 324
Ely 339
Soda Springs 316
Davis 257
Fresno 289
Santa Maria 346
Los Angeles WBO 327
Los Angeles WBAS 331
Riverside 367
La Jolla 362
Inyokern 412
Las Vegas 384
Page 382
Albuquerque 386
Phoenix 409
Tucson 391
El Paso 430
Midland 358
150
200
250
300
350
400

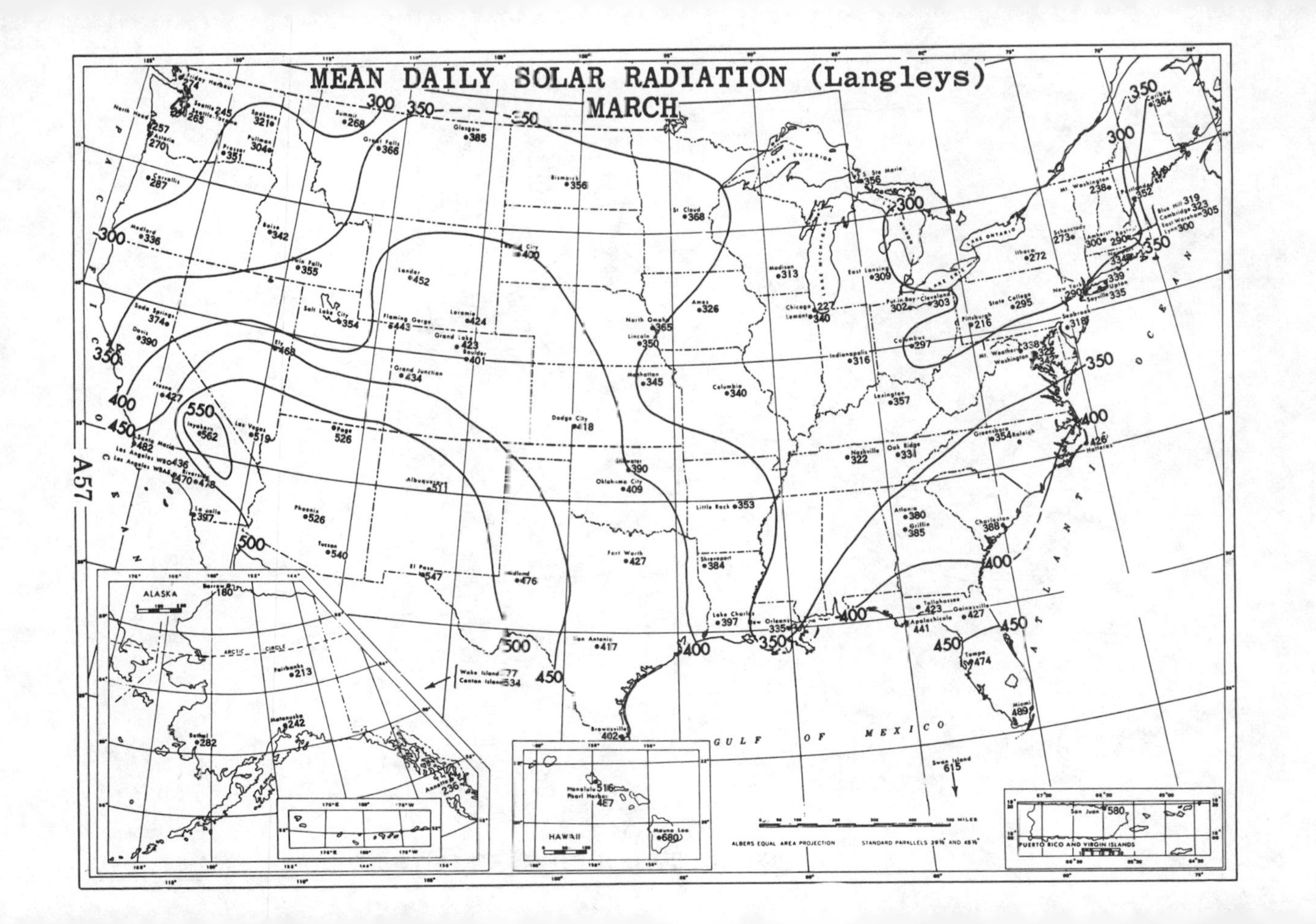
MEAN DAILY SOLAR RADIATION (Langleys)
MARCH
GULF OF MEXICO
ALASKA
HAWAII
PUERTO RICO AND VIRGIN ISLANDS
ARCTIC CIRCLE
ALBERS EQUAL AREA PROJECTION
STANDARD PARALLELS 29½° AND 45½°
Wake Island 77
Canton Island 534
Swan Island 615

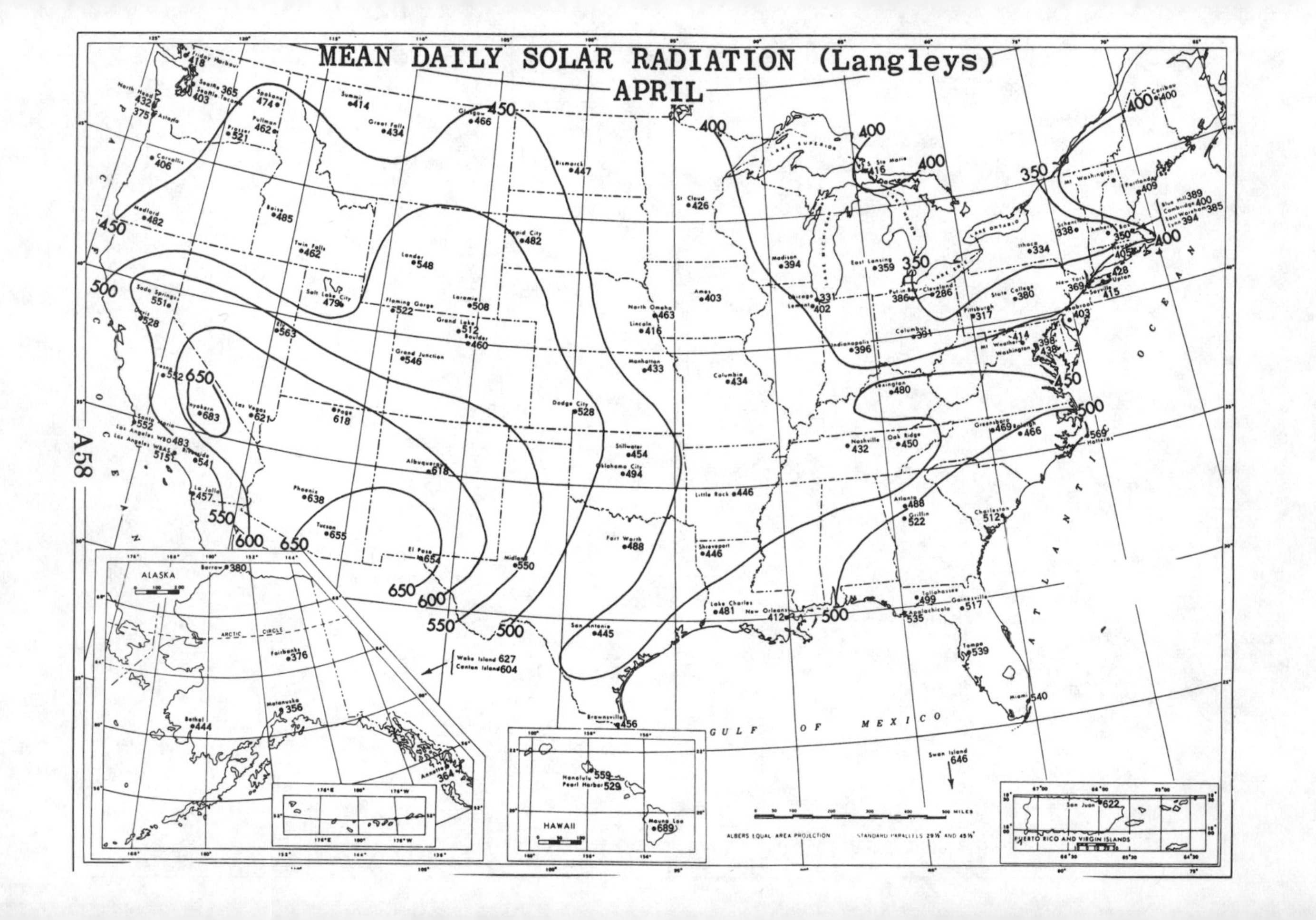
MEAN DAILY SOLAR RADIATION (Langleys)
APRIL
GULF OF MEXICO
ATLANTIC OCEAN
PACIFIC OCEAN
ALASKA
HAWAII
PUERTO RICO AND VIRGIN ISLANDS
ALBERS EQUAL AREA PROJECTION
STANDARD PARALLELS 29½° AND 45½°
Wake Island 627
Canton Island 604
Swan Island 646

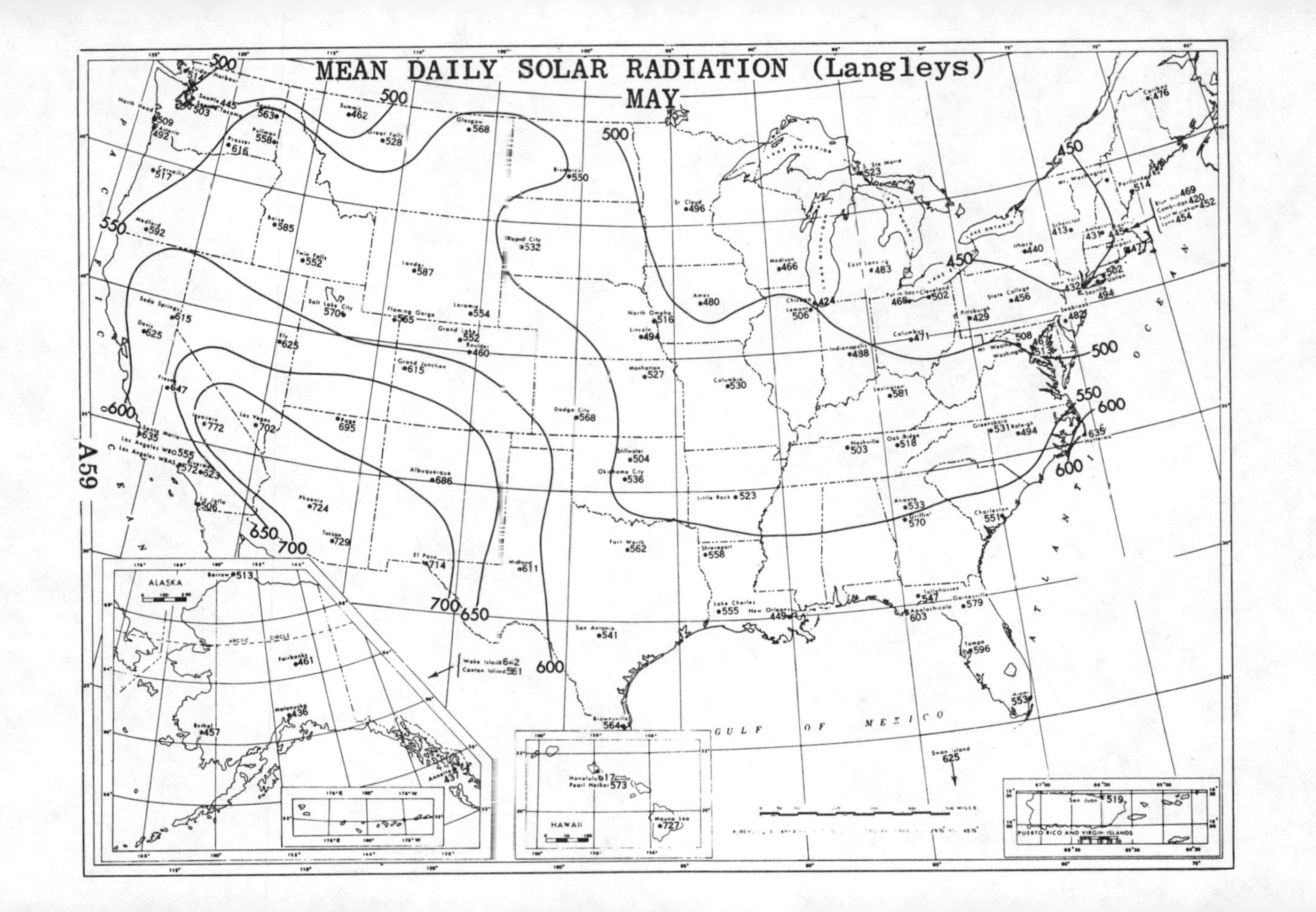

MEAN DAILY SOLAR RADIATION (Langleys)
MAY
GULF OF MEXICO
ALASKA
ARCTIC CIRCLE
HAWAII
PUERTO RICO AND VIRGIN ISLANDS
Wake Island
Canton Island
Swan Island 625

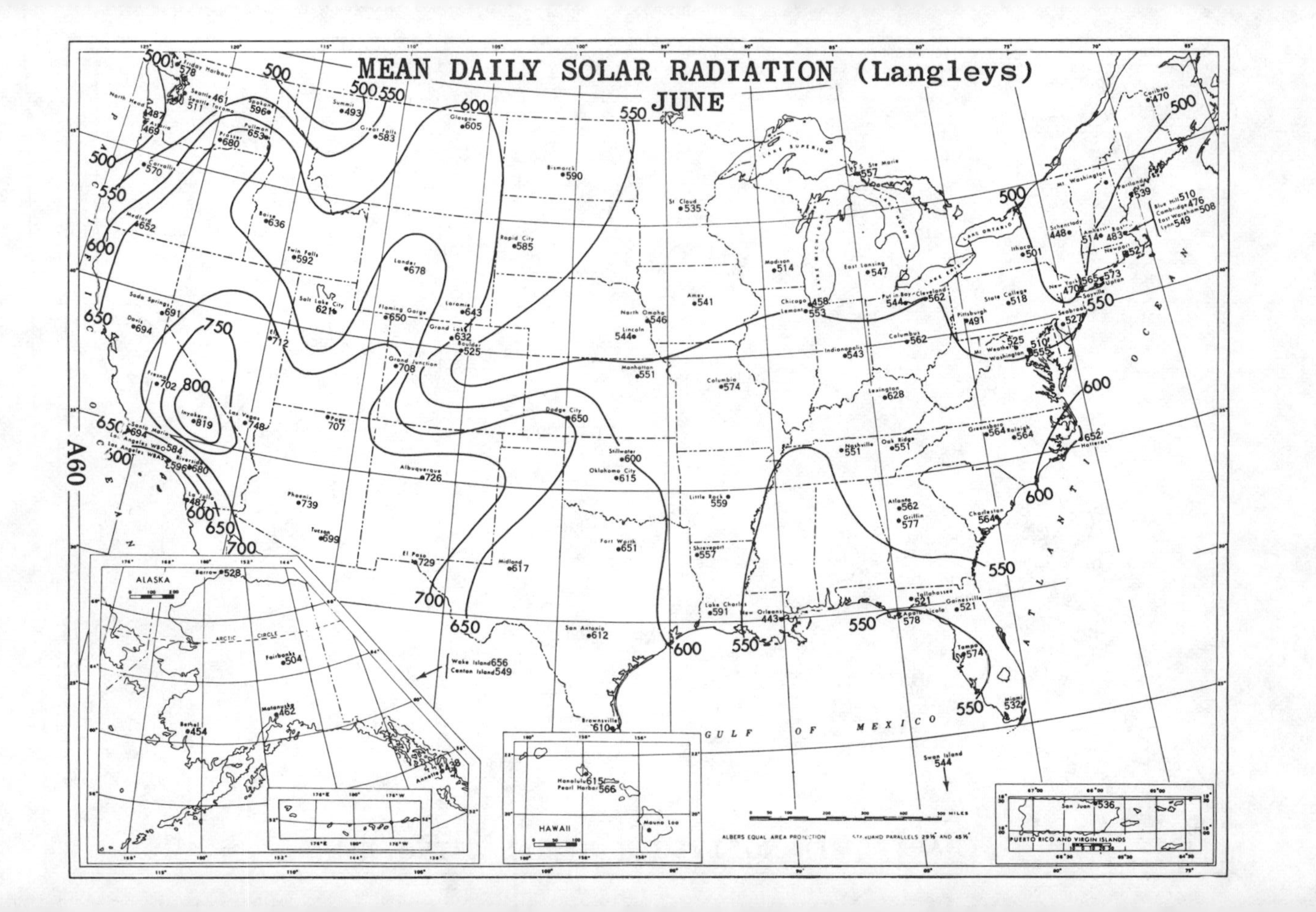
MEAN DAILY SOLAR RADIATION (Langleys)
JUNE
Seattle 461
Seattle Tacoma 511
Spokane 596
Summit 493
Great Falls 583
Glasgow 605
Bismarck 590
St Cloud 535
Sault Ste Marie 557
Caribou 470
Portland 539
Blue Hill 510
Cambridge 476
East Wareham 508
Lynn 549
Mt Washington
Schenectady 448
Newport 527
Ithaca 501
New York 470
Upton 573
Sayville
Seabrook 527
North Head 487
Astoria 469
Prosser 680
Pullman 653
Corvallis 570
Boise 636
Medford 652
Twin Falls 592
Rapid City 585
Lander 678
Salt Lake City 621
Laramie 643
Flaming Gorge 650
Grand Lake 632
Boulder 525
Grand Junction 708
Madison 514
East Lansing 547
Chicago 458
Lemont 553
Put in Bay 544
Cleveland 562
Pittsburgh 491
State College 518
North Omaha 546
Ames 541
Lincoln 544
Columbus 562
Indianapolis 543
Mt Weather 525
Washington 510
Sodo Springs 691
Davis 694
Ely 712
Fresno 702
Inyokern 819
Las Vegas 748
Page 707
Manhattan 551
Columbia 574
Lexington 628
Dodge City 650
Santa Maria 694
Los Angeles WBO 584
Los Angeles WBAS 596
Riverside 680
La Jolla 487
Albuquerque 726
Stillwater 600
Oklahoma City 615
Nashville 551
Oak Ridge 551
Greensboro 564
Raleigh 564
Hatteras 652
Phoenix 739
Tucson 699
Little Rock 559
Atlanta 562
Griffin 577
Charleston 564
El Paso 729
Midland 617
Fort Worth 651
Shreveport 557
Tallahassee 521
Gainesville 521
Lake Charles 591
New Orleans 443
Apalachicola 578
San Antonio 612
Tampa 574
Brownsville 610
Miami 532
GULF OF MEXICO
ATLANTIC OCEAN
PACIFIC OCEAN
Swan Island 544
Wake Island 656
Canton Island 549
ALASKA
Barrow 528
ARCTIC CIRCLE
Fairbanks 504
Matanuska 462
Bethel 454
Annette 438
HAWAII
Honolulu 615
Pearl Harbor 566
Mauna Loa
San Juan 536
PUERTO RICO AND VIRGIN ISLANDS
ALBERS EQUAL AREA PROJECTION
STANDARD PARALLELS 29½° AND 45½°
500 MILES

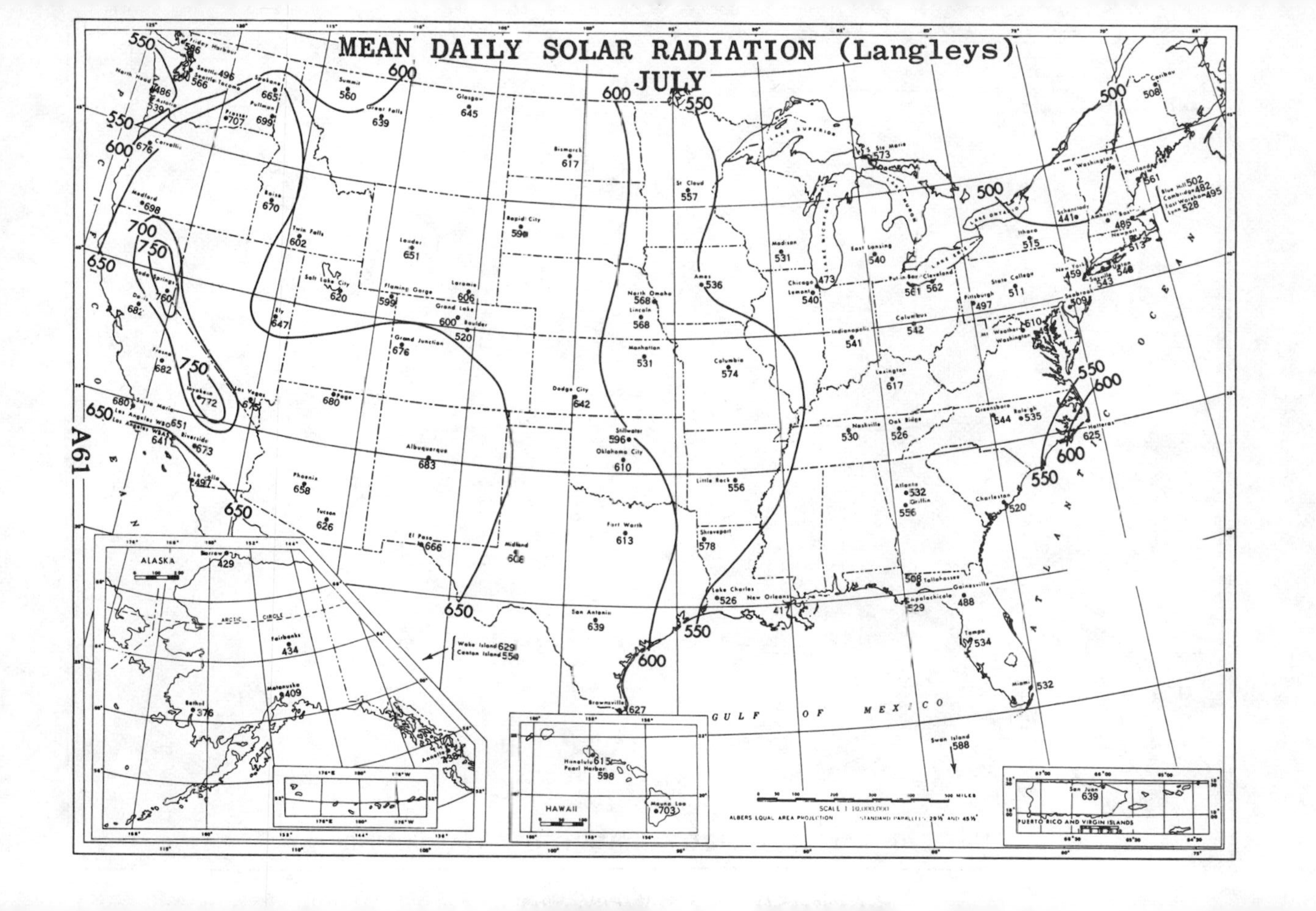
MEAN DAILY SOLAR RADIATION (Langleys)
JULY
GULF OF MEXICO
ALASKA
HAWAII
PUERTO RICO AND VIRGIN ISLANDS
ALBERS EQUAL AREA PROJECTION

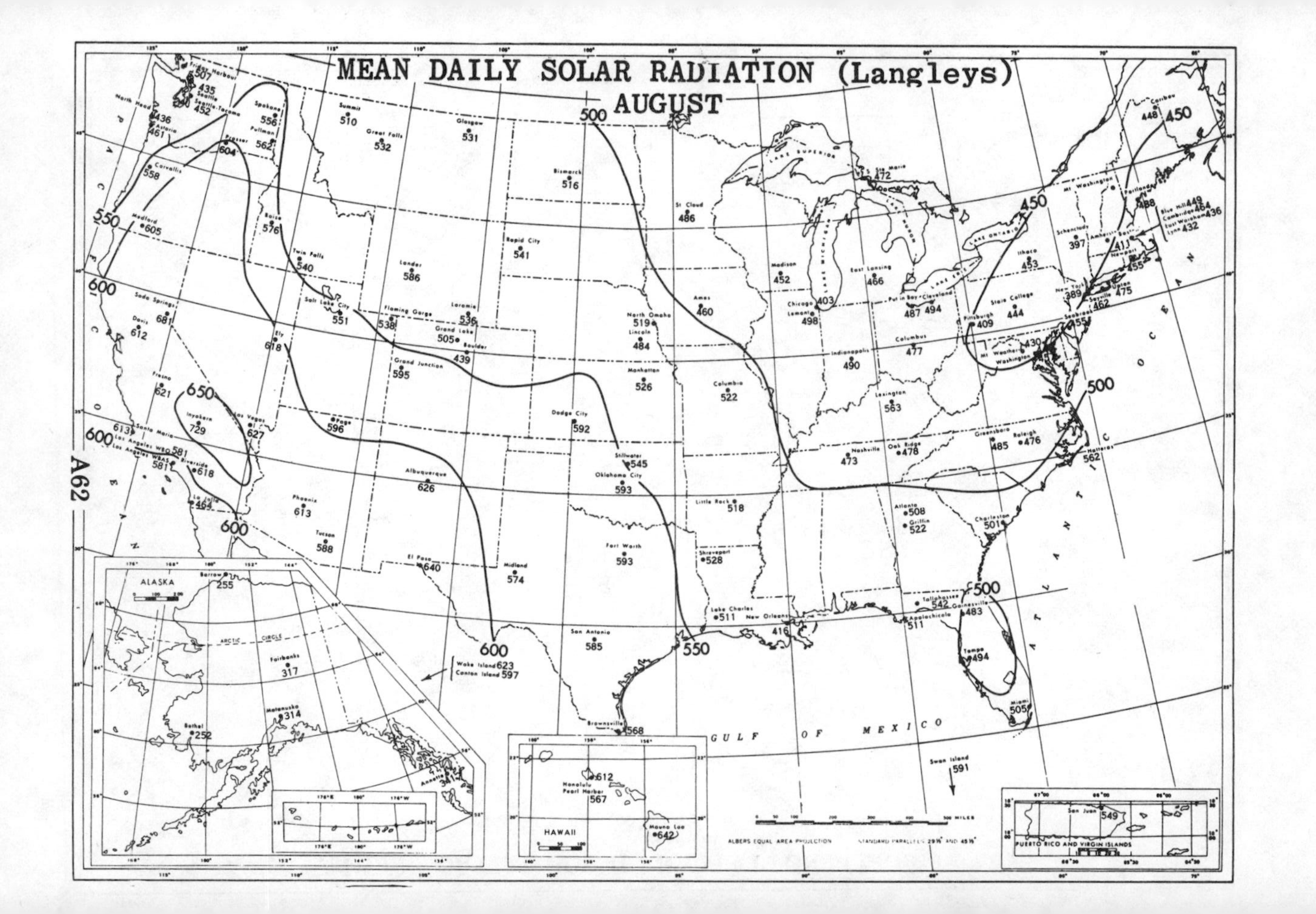

MEAN DAILY SOLAR RADIATION (Langleys)
AUGUST
Friday Harbour 507
Seattle 435
Seattle-Tacoma 452
North Head 436
Astoria 461
Spokane 556
Pullman 562
Prosser 604
Corvallis 558
Medford 605
Boise 576
Twin Falls 540
Summit 510
Great Falls 532
Glasgow 531
Bismarck 516
St Cloud 486
Rapid City 541
Lander 586
Salt Lake City 551
Flaming Gorge 538
Laramie 536
Grand Lake 505
Boulder 439
Grand Junction 595
Sacramento
Davis 612
Soda Springs 681
Ely 618
Fresno 621
Inyokern 729
Las Vegas 627
Santa Maria 613
Los Angeles WBO 581
Los Angeles WBAS 581
Riverside 618
La Jolla 464
Page 596
Phoenix 613
Tucson 588
Albuquerque 626
El Paso 640
Midland 574
Dodge City 592
North Omaha 519
Lincoln 484
Manhattan 526
Stillwater 545
Oklahoma City 593
Fort Worth 593
San Antonio 585
Brownsville 568
Ames 460
Columbia 522
Little Rock 518
Shreveport 528
Lake Charles 511
New Orleans 416
Madison 452
Chicago 403
Lemont 498
East Lansing 466
Sault Ste Marie 472
Indianapolis 490
Put in Bay 487
Cleveland 494
Columbus 477
Lexington 563
Nashville 473
Oak Ridge 478
Atlanta 508
Griffin 522
Tallahassee 542
Apalachicola 511
Gainesville 483
Tampa 494
Miami 505
Charleston 501
Greensboro 485
Raleigh 476
Hatteras 562
Pittsburgh 409
State College 444
Mt Weather 430
Washington
Seabrook 455
Ithaca 453
Schenectady 397
New York 389
Sayville 462
Upton 475
Newport 455
Amherst
Boston 411
Mt Washington
Portland 488
Caribou 448
Blue Hill 449
Cambridge 464
East Wareham 436
Lynn 432
Wake Island 623
Canton Island 597
Swan Island 591
LAKE SUPERIOR
LAKE MICHIGAN
LAKE HURON
LAKE ERIE
LAKE ONTARIO
PACIFIC OCEAN
ATLANTIC OCEAN
GULF OF MEXICO
ALASKA
Barrow 255
ARCTIC CIRCLE
Fairbanks 317
Matanuska 314
Bethel 252
Annette 341
HAWAII
Honolulu 612
Pearl Harbor 567
Mauna Loa 642
PUERTO RICO AND VIRGIN ISLANDS
San Juan 549
ALBERS EQUAL AREA PROJECTION
STANDARD PARALLELS 29½° AND 45½°
500 MILES

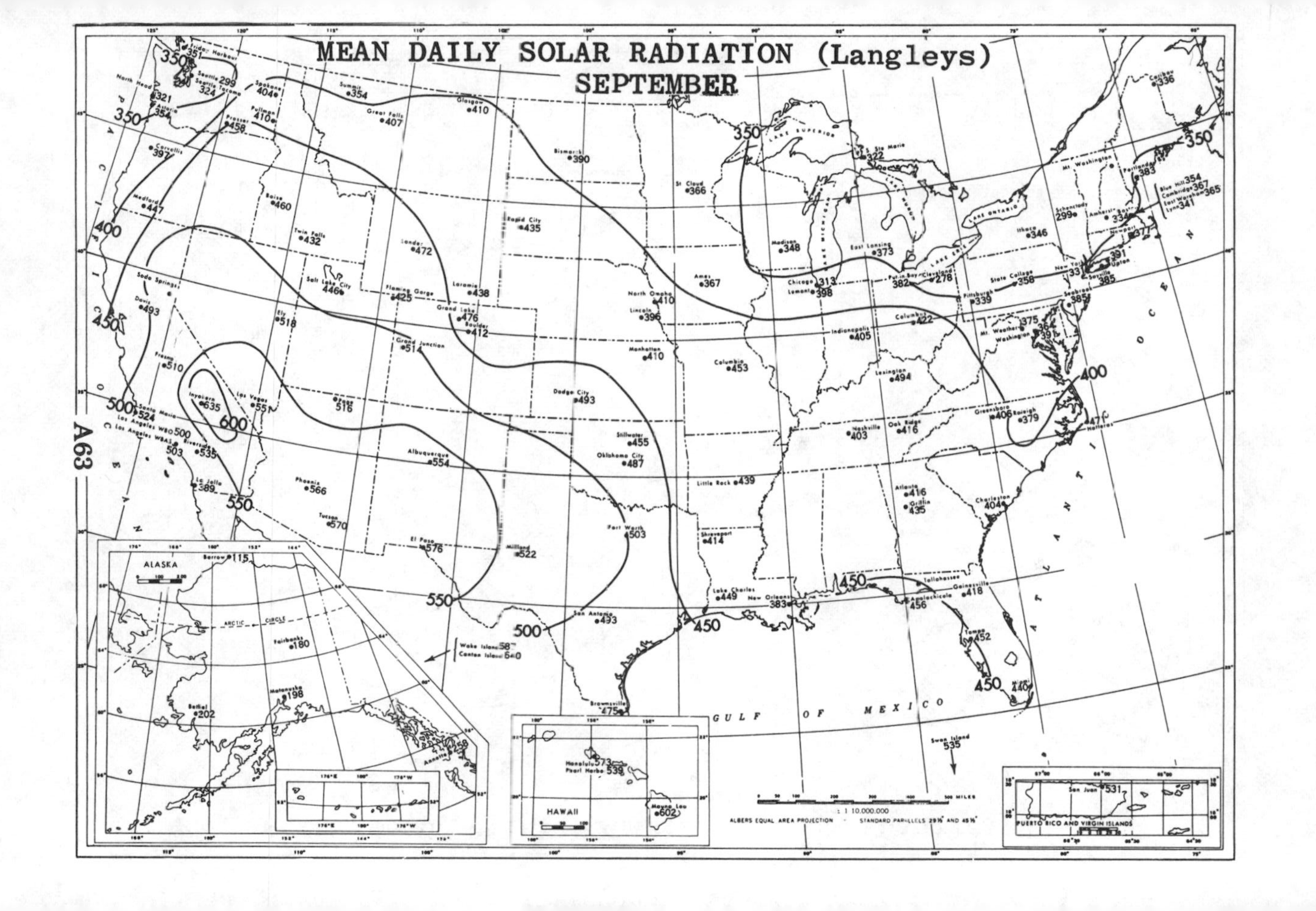
MEAN DAILY SOLAR RADIATION (Langleys)
SEPTEMBER
350
400
450
500
550
600
Glasgow 410
Great Falls 407
Summit 354
Spokane 404
Pullman 410
Prosser 458
Seattle 299
Seattle Tacoma 324
Astoria 354
North Head 321
Corvallis 397
Medford 447
Davis 493
Fresno 510
Santa Maria 524
Los Angeles WBO 500
Los Angeles WBAS 503
Riverside 535
La Jolla 389
Inyokern 635
Las Vegas 551
Ely 518
Boise 460
Twin Falls 432
Salt Lake City 446
Lander 472
Flaming Gorge 425
Grand Junction 514
Grand Lake 476
Boulder 412
Laramie 438
Rapid City 435
Bismarck 390
Phoenix 566
Tucson 570
Albuquerque 554
El Paso 576
Midland 522
Fort Worth 503
San Antonio 493
Brownsville 475
Dodge City 493
Stillwater 455
Oklahoma City 487
Manhattan 410
Lincoln 396
North Omaha 410
Ames 367
St Cloud 366
Columbia 453
Little Rock 439
Shreveport 414
Lake Charles 449
New Orleans 383
Madison 348
Chicago 313
Lemont 398
East Lansing 373
Sault Ste Marie 322
Indianapolis 405
Columbus 422
Lexington 494
Nashville 403
Oak Ridge 416
Atlanta 416
Griffin 435
Apalachicola 456
Tallahassee
Gainesville 418
Tampa 452
Miami 440
Charleston 404
Greensboro 406
Raleigh 379
Hatteras 471
Put-in-Bay 382
Cleveland 278
Pittsburgh 339
State College 358
Ithaca 346
Schenectady 299
New York 331
Sayville 385
Upton 391
Seabrook 385
Mt. Weather 375
Washington 364
Mt. Washington
Portland 383
Caribou 336
Amherst 334
Newport 377
Blue Hill 354
Cambridge 367
East Wareham 365
Lynn 341
Wake Island 587
Canton Island 640
Swan Island 535
ALASKA
Barrow 115
Fairbanks 180
Matanuska 198
Bethel 202
Annette 258
ARCTIC CIRCLE
HAWAII
Honolulu 573
Pearl Harbor 539
Mauna Loa 602
PUERTO RICO AND VIRGIN ISLANDS
San Juan 531
GULF OF MEXICO
ATLANTIC OCEAN
PACIFIC OCEAN
1:10,000,000
ALBERS EQUAL AREA PROJECTION — STANDARD PARALLELS 29½° AND 45½°
500 MILES

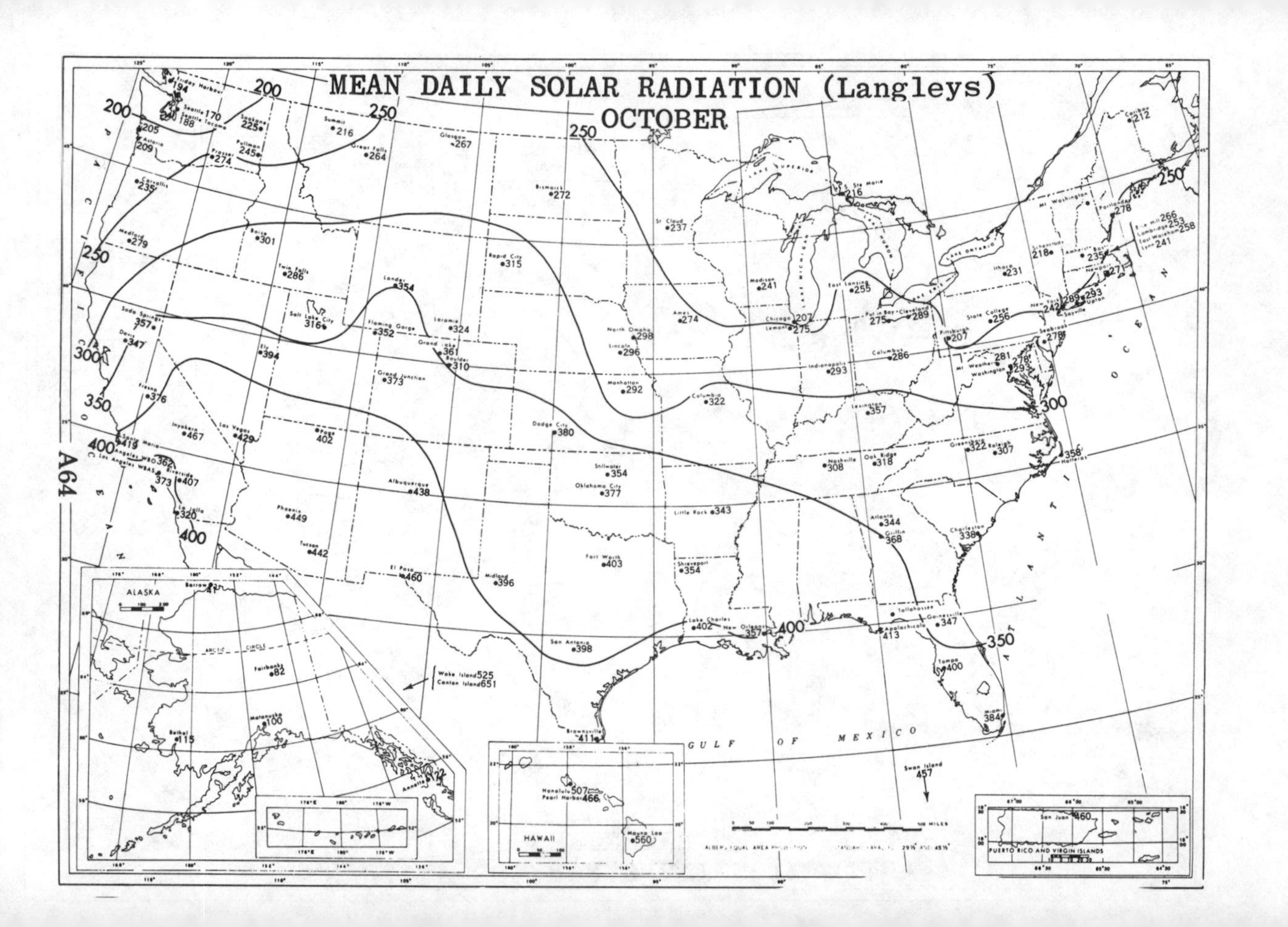
MEAN DAILY SOLAR RADIATION (Langleys)
OCTOBER
Friday Harbour 194
Seattle 170
Seattle Tacoma 188
205
Astoria 209
Spokane 225
Pullman 245
Prosser 274
Corvallis 235
Medford 279
Summit 216
Great Falls 264
Glasgow 267
Boise 301
Twin Falls 286
Lander 354
Salt Lake City 316
Soda Springs 357
Davis 347
Ely 394
Flaming Gorge 352
Laramie 324
Grand Lake 361
Boulder 310
Grand Junction 373
Fresno 376
Inyokern 467
Las Vegas 429
Page 402
Santa Maria 419
Los Angeles WBO 362
Los Angeles WBAS 373
Riverside 407
La Jolla 320
Phoenix 449
Tucson 442
Albuquerque 438
El Paso 460
Midland 396
Bismarck 272
Rapid City 315
St Cloud 237
North Omaha 298
Lincoln 296
Ames 274
Manhattan 292
Dodge City 380
Columbia 322
Stillwater 354
Oklahoma City 377
Little Rock 343
Fort Worth 403
Shreveport 354
San Antonio 398
Lake Charles 402
New Orleans 357
Brownsville 411
Madison 241
Chicago 207
Lemont 275
S. Ste Marie 216
East Lansing 255
Put in Bay 275
Cleveland 289
Indianapolis 293
Columbus 286
Lexington 357
Nashville 308
Oak Ridge 318
Atlanta 344
Griffin 368
Tallahassee
Apalachicola 413
Gainesville 347
Tampa 400
Miami 384
Charleston 338
Greensboro 322
Raleigh 307
Hatteras 358
Pittsburgh 207
State College 256
Mt Weather 281
Washington 278 293
Seabrook 278
Ithaca 231
Schenectady 218
New York 242
289
Sayville 293
Upton
Amherst
Boston 235
Newport 271
Mt Washington
Portland 278
Caribou 212
Blue Hill 266
Cambridge 253
East Wareham 258
Lynn 241
200
250
300
350
400
PACIFIC OCEAN
ATLANTIC OCEAN
GULF OF MEXICO
LAKE SUPERIOR
LAKE MICHIGAN
LAKE HURON
LAKE ONTARIO
ALASKA
Barrow 41
ARCTIC CIRCLE
Fairbanks 82
Matanuska 100
Bethel 115
Annette 122
Wake Island 525
Canton Island 651
HAWAII
Honolulu 507
Pearl Harbor 466
Mauna Loa 560
Swan Island 457
PUERTO RICO AND VIRGIN ISLANDS
San Juan 460
MILES

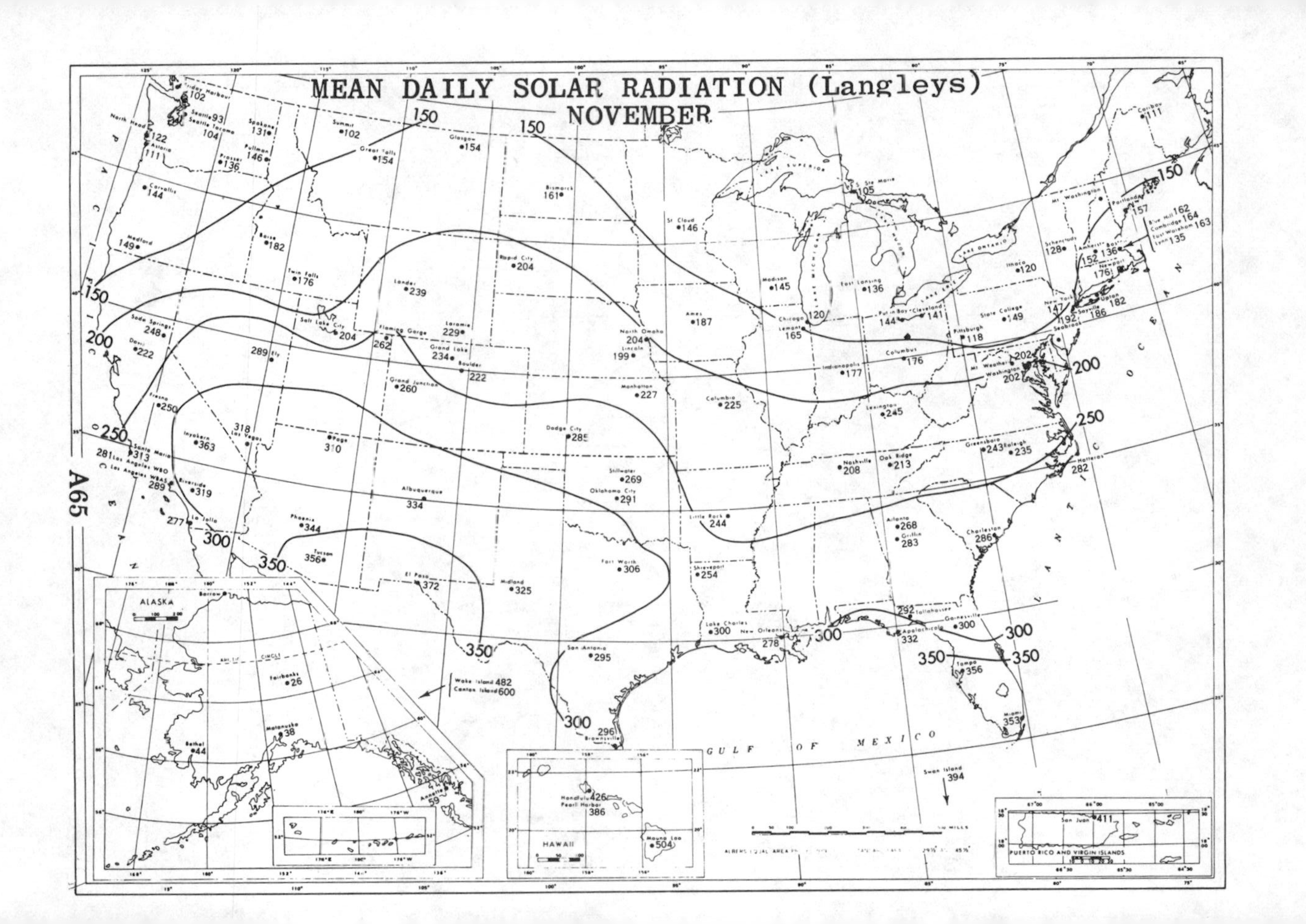
MEAN DAILY SOLAR RADIATION (Langleys)
NOVEMBER
150
200
250
300
350
Friday Harbour 102
Seattle 93
Seattle Tacoma 104
North Head 122
Astoria 111
Spokane 131
Pullman 146
Prosser 136
Summit 102
Great Falls 154
Glasgow 154
Bismarck 161
Corvallis 144
Medford 149
Boise 182
Twin Falls 176
Lander 239
Rapid City 204
St Cloud 146
Sault Ste Marie 105
Madison 145
East Lansing 136
Ithaca 120
Schenectady 128
Mt Washington
Caribou 111
Portland 157
Blue Hill 162
Cambridge 164
East Wareham 163
Lynn 135
Amherst 152
Boston 136
Newport 176
New York 147
Seabrook 192
Sayville 186
Upton 182
State College 149
Pittsburgh 118
Put-in-Bay 144
Cleveland 141
Chicago 120
Lemont 165
Ames 187
North Omaha 204
Lincoln 199
Salt Lake City 204
Flaming Gorge 262
Laramie 229
Grand Lake 234
Boulder 222
Soda Springs 248
Davis 222
Ely 289
Grand Junction 260
Manhattan 227
Columbia 225
Columbus 176
Indianapolis 177
Mt Weather 202
Washington 202
Lexington 245
Fresno 250
Las Vegas 318
Page 310
Dodge City 285
Inyokern 363
Santa Maria 313
Los Angeles WBO 281
Los Angeles WBAS 289
Riverside 319
La Jolla 277
Albuquerque 334
Stillwater 269
Oklahoma City 291
Nashville 208
Oak Ridge 213
Greensboro 243
Raleigh 235
Hatteras 282
Phoenix 344
Tucson 356
Little Rock 244
Atlanta 268
Griffin 283
Charleston 286
El Paso 372
Midland 325
Fort Worth 306
Shreveport 254
Lake Charles 300
New Orleans 278
Tallahassee 292
Apalachicola 332
Gainesville 300
Tampa 356
Miami 353
San Antonio 295
Brownsville 296
Wake Island 482
Canton Island 600
GULF OF MEXICO
Swan Island 394
ALASKA
Barrow
Fairbanks 26
Matanuska 38
Bethel 44
Annette 59
HAWAII
Honolulu 426
Pearl Harbor 386
Mauna Loa 504
San Juan 411
PUERTO RICO AND VIRGIN ISLANDS

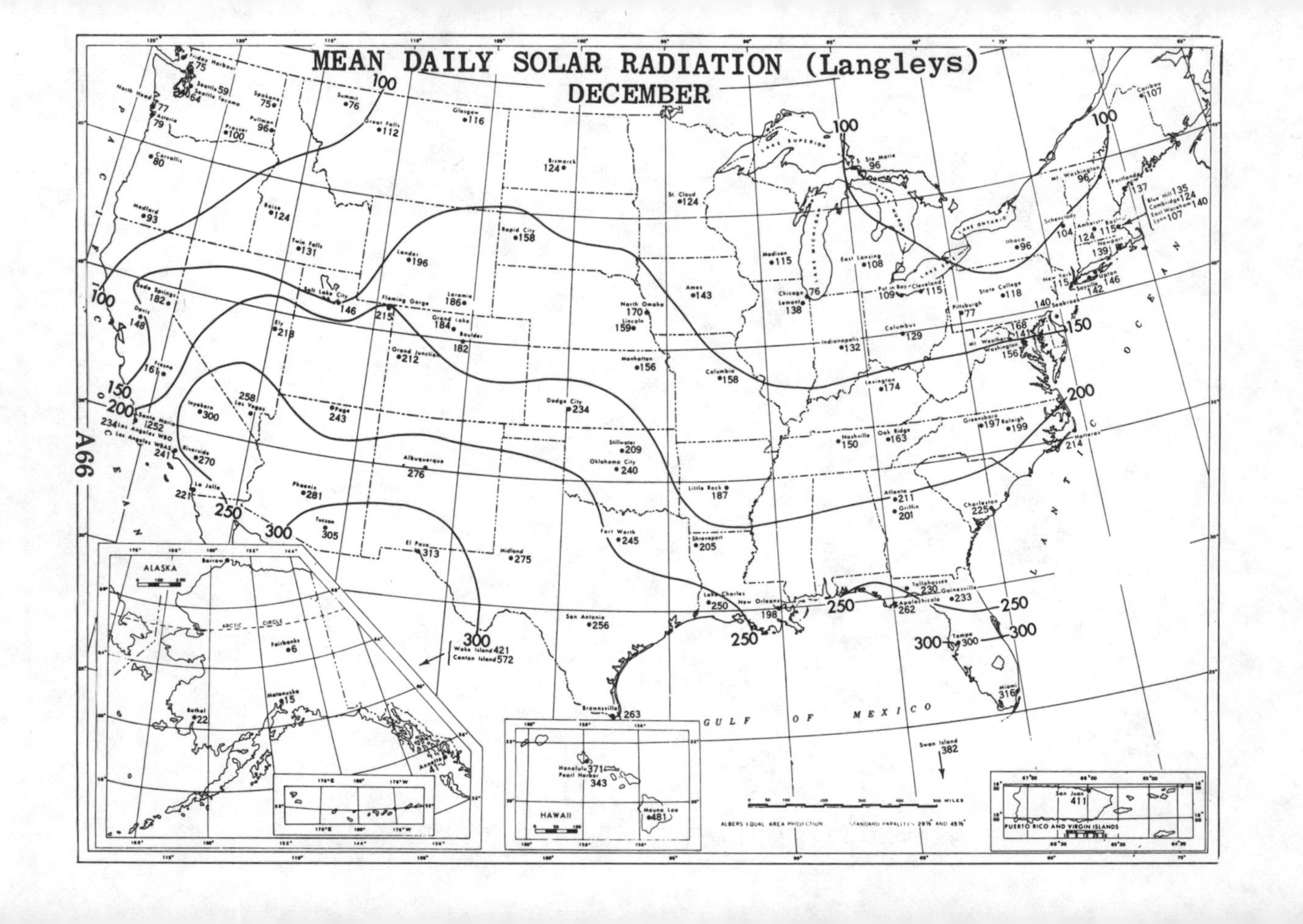
MEAN DAILY SOLAR RADIATION (Langleys)
DECEMBER
Friday Harbour 75
Seattle 59
Seattle Tacoma 64
North Head 77
Astoria 79
Spokane 75
Pullman 96
Prosser 100
Summit 76
Great Falls 112
Glasgow 116
Corvallis 80
Medford 93
Boise 124
Twin Falls 131
Bismarck 124
St. Cloud 124
Rapid City 158
Lander 196
Salt Lake City 146
Flaming Gorge 215
Laramie 186
Grand Lake 184
Boulder 182
Grand Junction 212
Ely 218
Soda Springs 182
Davis 148
Fresno 161
Santa Maria 252
Los Angeles WBO 234
Los Angeles WBAS 241
Inyokern 300
Las Vegas 258
Page 243
Riverside 270
La Jolla 221
Phoenix 281
Tucson 305
Albuquerque 276
El Paso 313
Midland 275
North Omaha 170
Lincoln 159
Ames 143
Manhattan 156
Dodge City 234
Columbia 158
Stillwater 209
Oklahoma City 240
Little Rock 187
Fort Worth 245
Shreveport 205
San Antonio 256
Brownsville 263
Lake Charles 250
New Orleans 198
Madison 115
Chicago 76
Lemont 138
East Lansing 108
Sault Ste Marie 96
Put in Bay 109
Cleveland 115
Indianapolis 132
Columbus 129
Lexington 174
Nashville 150
Oak Ridge 163
Atlanta 211
Griffin 201
Tallahassee 230
Apalachicola 262
Gainesville 233
Tampa 300
Miami 316
Charleston 225
Greensboro 197
Raleigh 199
Hatteras 214
Pittsburgh 77
State College 118
Washington 156
Mt Weather 168
Seabrook 140
New York 115
Sayville 142
Upton 146
Ithaca 96
Schenectady 104
Amherst 124
Boston 115
Newport 139
Mt Washington 96
Portland 137
Caribou 107
Blue Hill 135
Cambridge 124
East Wareham 140
Lynn 107
Wake Island 421
Canton Island 572
Swan Island 382
LAKE SUPERIOR
LAKE MICHIGAN
LAKE HURON
LAKE ONTARIO
LAKE ERIE
PACIFIC OCEAN
ATLANTIC OCEAN
GULF OF MEXICO
ALASKA
Barrow
ARCTIC CIRCLE
Fairbanks 6
Matanuska 15
Bethel 22
Annette 41
HAWAII
Honolulu 371
Pearl Harbor 343
Mauna Loa 481
PUERTO RICO AND VIRGIN ISLANDS
San Juan 411
ALBERS EQUAL AREA PROJECTION
STANDARD PARALLELS 29½° AND 45½°
MILES
100
150
200
250
300

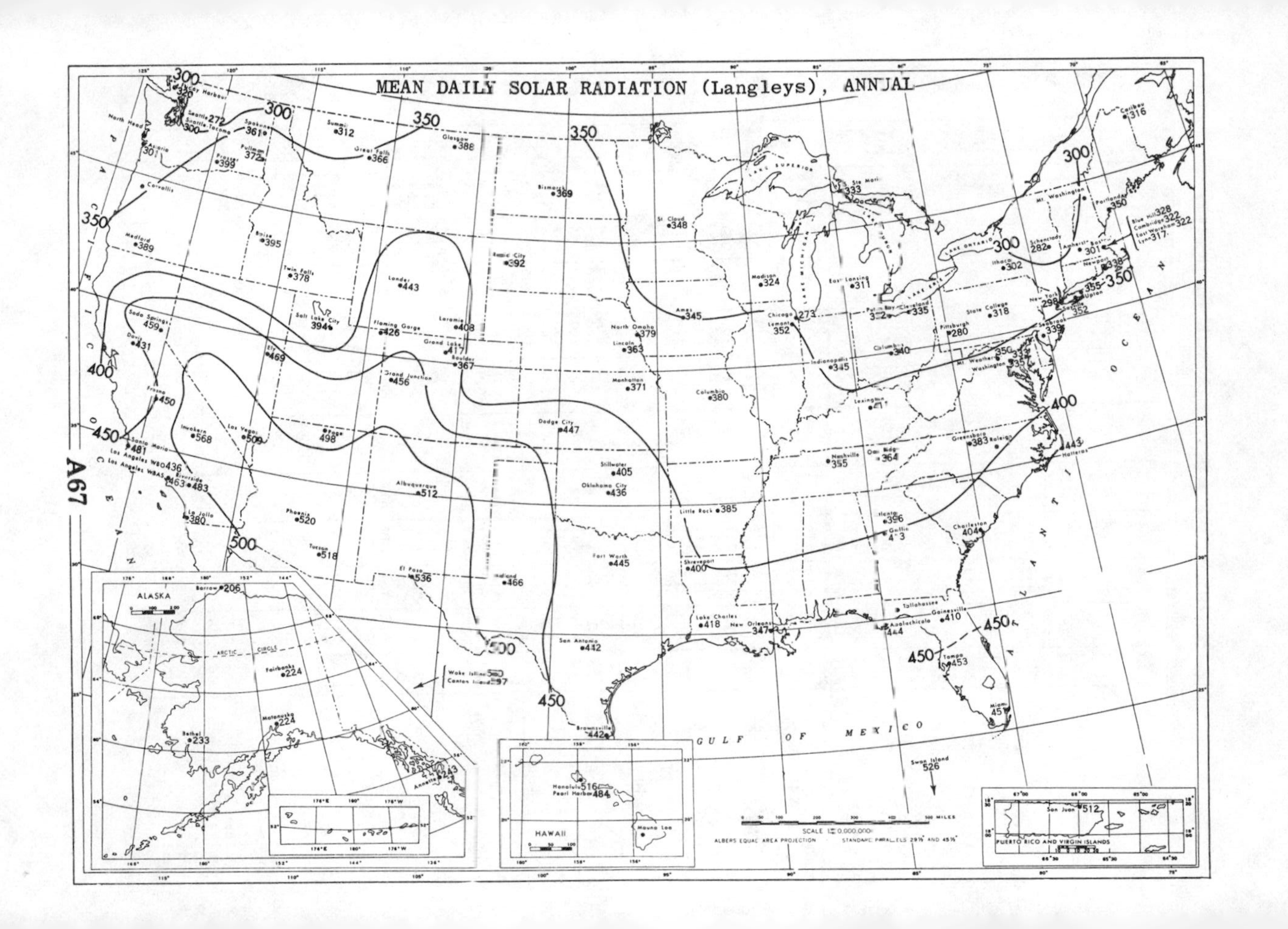
MEAN DAILY SOLAR RADIATION (Langleys), ANNUAL
Seattle 272
Seattle-Tacoma 300
Astoria 301
North Head
Corvallis
Medford 389
Spokane 361
Pullman 372
Prosser 399
Summit 312
Great Falls 366
Glasgow 388
Bismarck 369
Boise 395
Twin Falls 378
Lander 443
Salt Lake City 394
Flaming Gorge 426
Laramie 408
Rapid City 392
Grand Lake 417
Boulder 367
Grand Junction 456
Ely 469
Soda Springs 459
Davis 431
Fresno 450
Inyokern 568
Las Vegas 509
Page 498
Santa Maria 481
Los Angeles WBO 436
Los Angeles WBAS 463
Riverside 483
La Jolla 380
Phoenix 520
Tucson 518
Albuquerque 512
El Paso 536
Midland 466
Dodge City 447
St Cloud 348
Ames 345
North Omaha 379
Lincoln 363
Manhattan 371
Columbia 380
Stillwater 405
Oklahoma City 436
Fort Worth 445
Little Rock 385
Shreveport 400
Lake Charles 418
New Orleans 347
San Antonio 442
Brownsville 442
Madison 324
Chicago 273
Lemont 352
East Lansing 311
Sault Ste Marie 333
Indianapolis 345
Columbus 340
Cleveland 335
Pittsburgh 280
Nashville 355
Oak Ridge 364
Atlanta 396
Charleston 404
Greensboro 383
Raleigh
Hatteras 443
Tallahassee
Apalachicola 444
Gainesville 410
Tampa 453
Miami 451
Ithaca 302
Schenectady 282
State College 318
New York 298
Upton 355
Seabrook 339
Mt Weather 350
Washington
Mt Washington
Portland 350
Caribou 316
Amherst
Boston 301
Newport 338
Blue Hill 328
Cambridge 322
East Wareham 322
Lynn 317
Wake Island
Canton Island
Swan Island 526
LAKE SUPERIOR
LAKE MICHIGAN
LAKE ONTARIO
LAKE ERIE
GULF OF MEXICO
PACIFIC OCEAN
ATLANTIC OCEAN
ALASKA
Barrow 206
ARCTIC CIRCLE
Fairbanks 224
Matanuska 224
Bethel 233
Annette
HAWAII
Honolulu 516
Pearl Harbor 484
Mauna Loa
PUERTO RICO AND VIRGIN ISLANDS
San Juan 512
SCALE 1:10,000,000
ALBERS EQUAL AREA PROJECTION
STANDARD PARALLELS 29½° AND 45½°
300
350
400
450
500

MEAN DAILY SOLAR RADIATION (Langleys)

STATES AND STATIONS	JAN	FEB	MAR	APR	MAY	JUNE	JULY	AUG	SEPT	OCT	NOV	DEC	ANNUAL
ALASKA, Annette	63	115	236	364	437	438	438	341	258	122	59	41	243
Barrow	#	38	180	380	513	528	429	255	115	41	#	#	206
Bethel	38	108	282	444	457	454	376	252	202	115	44	22	233
Fairbanks	16	71	213	376	461	504	434	317	180	82	26	6	224
Matanuska	32	92	242	356	436	462	409	314	198	100	38	15	224
ARIZ., Page	300	382	526	618	695	707	680	596	516	402	310	243	498
Phoenix	301	409	526	638	724	739	658	613	566	449	344	281	520
Tucson	315	391	540	655	729	699	626	588	570	442	356	305	518
ARK., Little Rock	188	260	353	446	523	559	556	518	439	343	244	187	385
CALIFORNIA, Davis	174	257	390	528	625	694	682	612	493	347	222	148	431
Fresno	184	289	427	552	647	702	682	621	510	376	250	161	450
Inyokern (China Lake)	306	412	562	683	772	819	772	729	635	467	363	300	568
LaJolla	244	302	397	457	506	487	497	464	389	320	277	221	380
Los Angeles WBAS	248	331	470	515	572	596	641	581	503	373	289	241	463
Los Angeles WBO	243	327	436	483	555	584	651	581	500	362	281	234	436
Riverside ‡	275	367	478	541	623	680	673	618	535	407	319	270	483
Santa Maria	263	346	482	552	635	694	680	613	524	419	313	252	481
Soda Springs	223	316	374	551	615	691	760	681	510	357	248	182	459
COLO., Boulder	201	268	401	460	460	525	520	439	412	310	222	182	367
Grand Junction	227	324	434	546	615	708	676	595	514	373	260	212	456
Grand Lake (Granby)	212	313	423	512	552	632	600	505	476	361	234	184	417
D. C., Washington (C.O.)	174	266	344	411	551	494	536	446	375	299	211	166	356
American University	158	231	322	398	467	510	496	440	364	278	192	141	333
Silver Hill	177	247	342	438	513	555	511	457	391	293	202	156	357
FLA., Apalachicola	298	367	441	535	603	578	529	511	456	413	332	262	444
Belle Isle	297	330	412	463	483	464	488	461	400	366	313	291	397
Gainesville	267	343	427	517	579	521	488	483	418	347	300	233	410
Miami Airport	349	415	489	540	553	532	532	505	440	384	353	316	451
Tallahassee	274	311	423	499	547	521	508	542	*	*	292	230	---
Tampa	327	391	474	539	596	574	534	494	452	400	356	300	453
GA., Atlanta	218	290	380	488	533	562	532	508	416	344	268	211	396
Griffin	234	295	385	522	570	577	556	522	435	368	283	201	413
HAWAII, Honolulu	363	422	516	559	617	615	615	612	573	507	426	371	516
Mauna Loa Obs.	522	576	680	689	727	*	703	642	602	560	504	481	---
Pearl Harbor	359	400	487	529	573	566	598	567	539	466	386	343	484
IDAHO, Boise	138	236	342	485	585	636	670	576	460	301	182	124	395
Twin Falls	163	240	355	462	552	592	602	540	432	286	176	131	378
ILL., Chicago	96	147	227	331	424	458	473	403	313	207	120	76	273
Lemont	170	242	340	402	506	553	540	498	398	275	165	138	352
IND., Indianapolis	144	213	316	396	488	543	541	490	405	293	177	132	345
IOWA, Ames	174	253	326	403	480	541	436	460	367	274	187	143	345
KANS., Dodge City	255	316	418	528	568	650	642	592	493	380	285	234	447
Manhattan	192	264	345	433	527	551	531	526	410	292	227	156	371
KY., Lexington	172	263	357	480	581	628	617	563	494	357	245	174	411
LA., Lake Charles	245	306	397	481	555	591	526	511	449	402	300	250	418
New Orleans	214	259	335	412	449	443	417	416	383	357	278	198	347
Shreveport	232	292	384	446	558	557	578	528	414	354	254	205	400
MAINE, Caribou	133	231	364	400	476	470	508	448	336	212	111	107	316
Portland	152	235	352	409	514	539	561	488	383	278	157	137	350
MASS., Amherst	116	*	300	*	431	514	*	---	---	---	152	124	---
Blue Hill	153	228	319	389	469	510	502	449	354	266	162	135	328
Boston	129	194	290	350	445	483	486	411	334	235	136	115	301
Cambridge	153	235	323	400	420	476	482	464	367	253	164	124	322
East Wareham	140	218	305	385	452	508	495	436	365	258	163	140	322
Lynn	118	209	300	394	454	549	528	432	341	241	135	107	317
MICH., East Lansing	121	210	309	359	483	547	540	466	373	255	136	108	311
Sault Ste. Marie	130	225	356	416	523	557	573	472	322	216	105	96	333
MINN., St. Cloud	168	260	368	426	496	535	557	486	366	237	146	124	348
MO., Columbia (C. O.)	173	251	340	434	530	574	574	522	453	322	225	158	380
University of Missouri	166	248	324	429	501	560	583	509	417	324	177	146	365
MONT., Glasgow	154	258	385	466	568	605	645	531	410	267	154	116	388
Great Falls	140	232	366	434	528	583	639	532	407	264	154	112	366
Summit	122	162	268	414	462	493	560	510	354	216	102	76	312
NEBR., Lincoln	188	259	350	416	494	544	568	484	396	296	199	159	363
North Omaha	193	299	365	463	516	546	568	519	410	298	204	170	379

MEAN DAILY SOLAR RADIATION (Langleys)

STATES AND STATIONS	JAN	FEB	MAR	APR	MAY	JUNE	JULY	AUG	SEPT	OCT	NOV	DEC	ANNUAL
NEV., Ely	236	339	468	563	625	712	647	618	518	394	289	218	469
Las Vegas	277	384	519	621	702	748	675	627	551	429	318	258	509
N. J., Seabrook	157	227	318	403	482	527	509	455	385	278	192	140	339
N. H., Mt. Washington	117	218	238	*	*	*	---	---	*	*	*	96	---
N. Mex., Albuquerque	303	386	511	618	686	726	683	626	554	438	334	276	512
N. Y., Ithaca	116	194	272	334	440	501	515	453	346	231	120	96	302
N. Y. Central Park	130	199	290	369	432	470	459	389	331	242	147	115	298
Sayville	160	249	335	415	494	565	543	462	385	289	186	142	352
Schenectady	130	200	273	338	413	448	441	397	299	218	128	104	282
Upton	155	232	339	428	502	573	543	475	391	293	182	146	355
N. C., Greensboro	200	276	354	469	531	564	544	485	406	322	243	197	383
Hatteras	238	317	426	569	635	652	625	562	471	358	282	214	443
Raleigh	235	302	*	466	494	564	535	476	379	307	235	199	---
N. D., Bismarck	157	250	356	447	550	590	617	516	390	272	161	124	369
OHIO., Cleveland	125	183	303	286	502	562	562	494	278	289	141	115	335
Columbus	128	200	297	391	471	562	542	477	422	286	176	129	340
Put-in-Bay	126	204	302	386	468	544	561	487	382	275	144	109	332
OKLA., Oklahoma City	251	310	409	494	536	615	610	593	487	377	291	240	436
Stillwater	205	289	390	454	504	600	596	546	455	354	269	209	405
OREG., Astoria	90	162	270	375	492	469	539	461	354	209	111	79	301
Corvallis	89	*	287	406	517	570	676	558	397	235	144	80	---
Medford	116	215	336	482	592	652	698	605	447	279	149	93	389
PA., Pittsburgh	94	169	216	317	429	491	497	409	339	207	118	77	280
State College	133	201	295	380	456	518	511	444	358	256	149	118	318
R. I., Newport	155	232	334	405	477	527	513	455	377	271	176	139	338
S. C., Charleston	252	314	388	512	551	564	520	501	404	338	286	225	404
S. D., Rapid City	183	277	400	482	532	585	590	541	435	315	204	158	392
TENN., Nashville	149	228	322	432	503	551	530	473	403	308	208	150	355
Oak Ridge	161	239	331	450	518	551	526	478	416	318	213	163	364
TEXAS, Brownsville	297	341	402	456	564	610	627	568	475	411	296	263	442
El Paso	333	430	547	654	714	729	666	640	576	460	372	313	536
Ft. Worth	250	320	427	488	562	651	613	593	503	403	306	245	445
Midland	283	358	476	550	611	617	608	574	522	396	325	275	466
San Antonio	279	347	417	445	541	612	639	585	493	398	295	256	442
UTAH, Flaming Gorge	238	298	443	522	565	650	599	538	425	352	262	215	426
Salt Lake City	163	256	354	479	570	621	620	551	446	316	204	146	394
VA., Mt. Weather	172	274	338	414	508	525	510	430	375	281	202	168	350
WASH., North Head	*	167	257	432	509	487	486	436	321	205	122	77	---
Friday Harbor	87	157	274	418	514	578	586	507	351	194	102	75	320
Prosser	117	222	351	521	616	680	707	604	458	274	136	100	399
Pullman	121	205	304	462	558	653	699	562	410	245	146	96	372
University of Washington	67	126	245	364	445	461	496	435	299	170	93	59	272
Seattle Tacoma	75	139	265	403	503	511	566	452	324	188	104	64	300
Spokane	119	204	321	474	563	596	665	556	404	225	131	75	[illegible]
WIS., Madison †	148	220	313	394	466	514	531	452	348	241	145	115	324
WYO., Lander	226	324	452	548	587	678	651	586	472	354	239	196	445
Laramie	216	295	424	508	554	643	606	536	438	324	229	186	408
ISLAND STATIONS													
Canton Island	588	626	634	604	561	549	550	597	640	651	600	572	597
San Juan, P. R.	404	481	580	622	519	536	639	549	531	460	411	411	512
Swan Island	442	496	615	646	625	544	588	591	535	457	394	382	526
Wake Island	438	518	577	627	642	656	629	623	587	525	482	421	560

NOTES:
- * Denotes only one year of data for the month -- no means computed.
- --- No data for the month (or incomplete data for the year).
- # Barrow is in darkness during the winter months.
- † Madison data after 1957 not used due to exposure influences.
- ‡ Riverside data prior to March 1952 not used-instrumental discrepancies.

These charts and table are based on all usable solar radiation data, direct and diffuse, measured on a horizontal surface and published in the Monthly Weather Review and Climatological Data National Summary through 1962. All data were measured in, or were reduced to, the International Scale of Pyrheliometry, 1956.

Langley is the unit used to note one gram calorie per square centimeter.

MEAN SKY COVER, SUNRISE TO SUNSET, (In Tenths), JANUARY

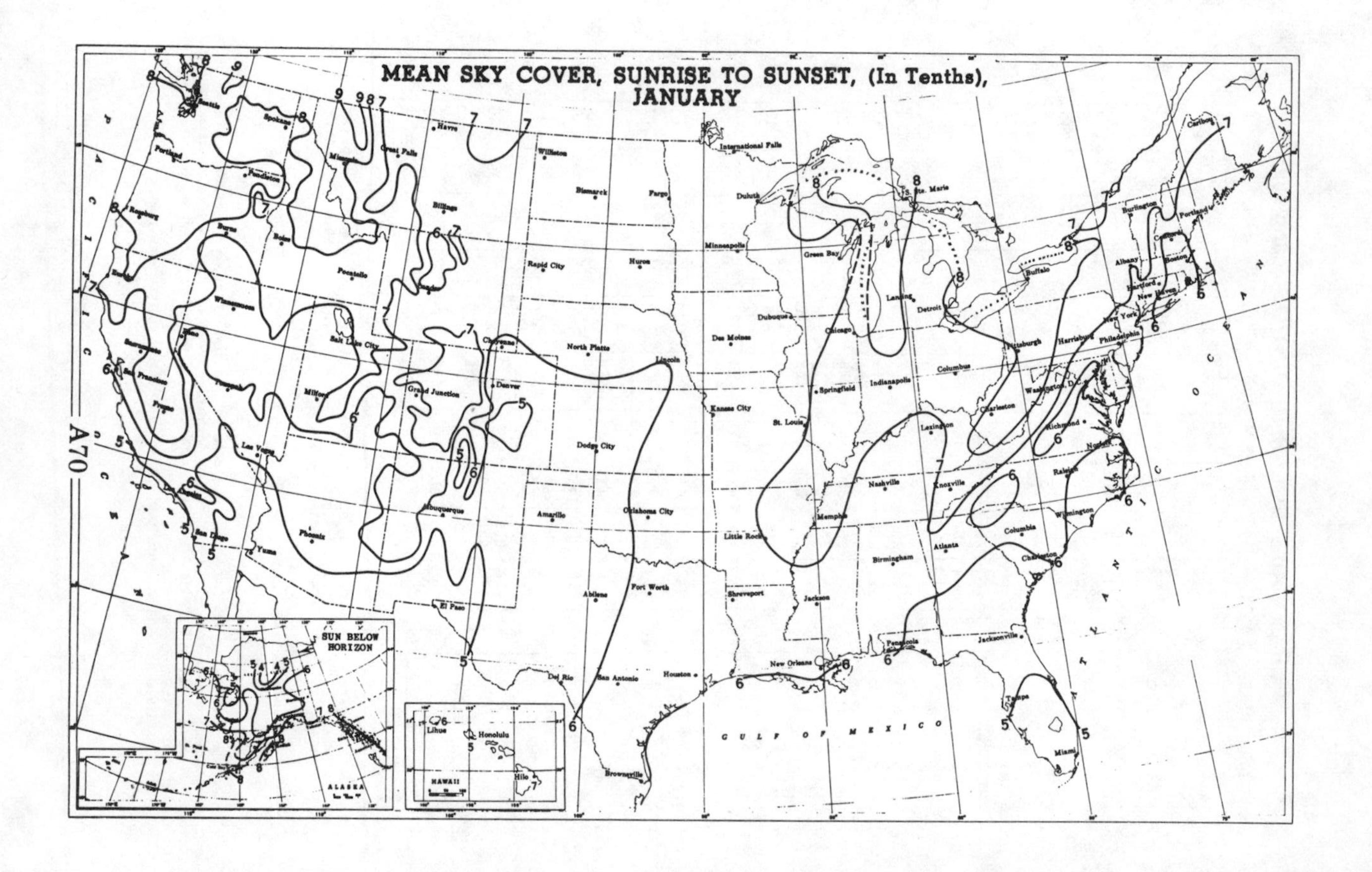

MEAN SKY COVER, SUNRISE TO SUNSET, (In Tenths), FEBRUARY

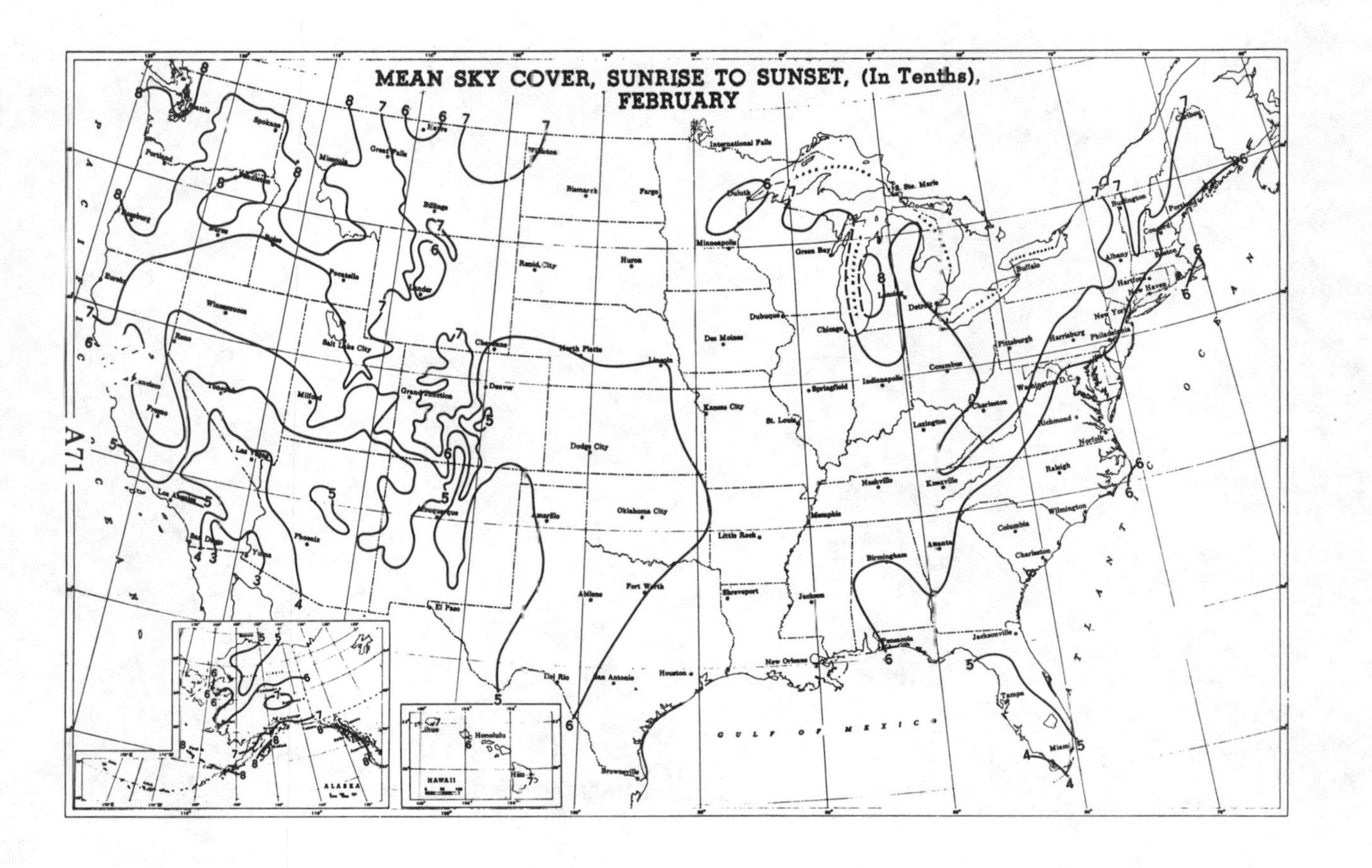

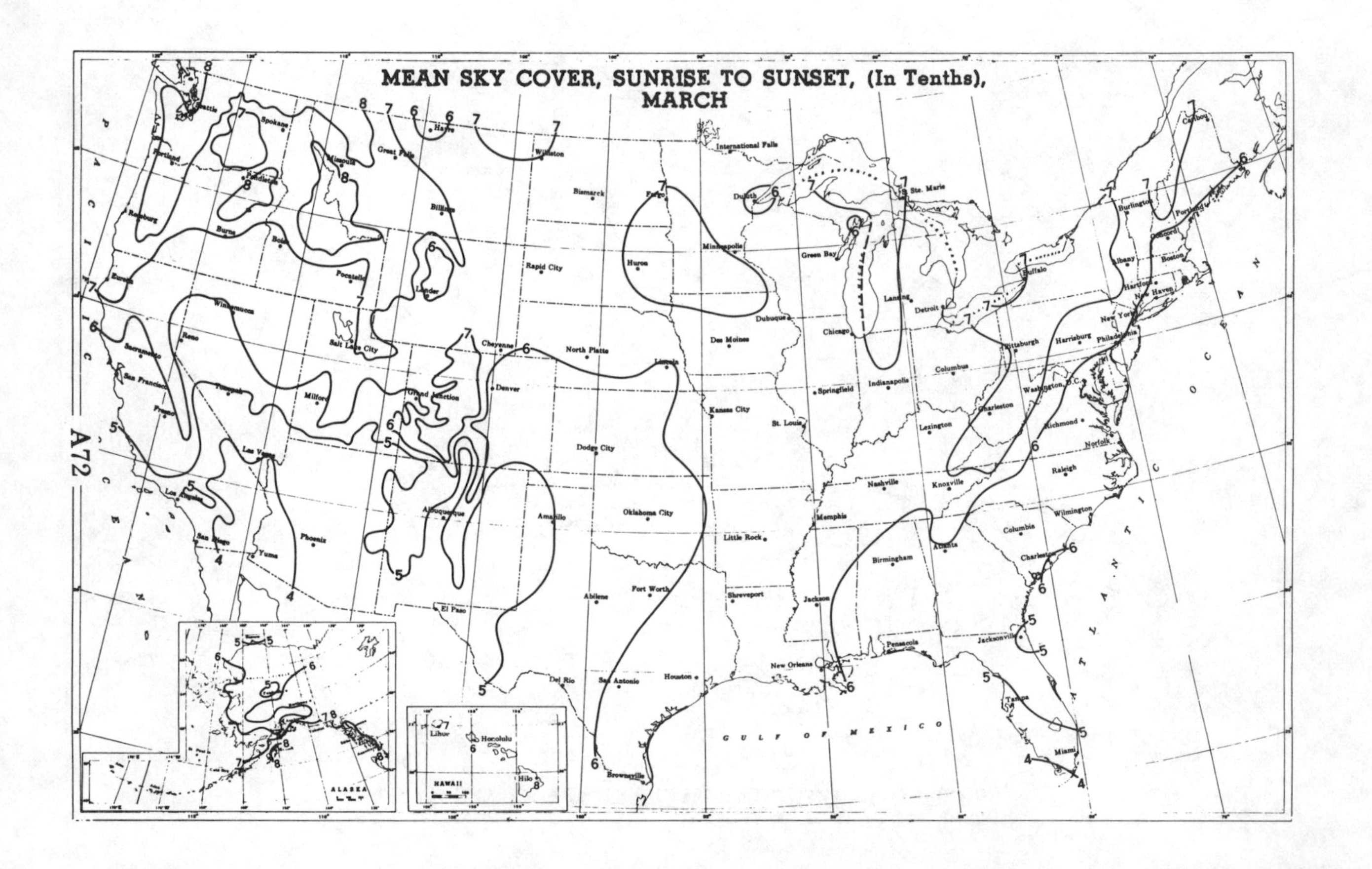
MEAN SKY COVER, SUNRISE TO SUNSET, (In Tenths),
MARCH
PACIFIC OCEAN
ATLANTIC OCEAN
GULF OF MEXICO
ALASKA
HAWAII
Lihue
Honolulu
Hilo
Seattle
Portland
Roseburg
Eureka
Spokane
Pendleton
Burns
Boise
Winnemucca
Reno
Sacramento
San Francisco
Fresno
Tonopah
Los Angeles
San Diego
Las Vegas
Yuma
Phoenix
Milford
Salt Lake City
Pocatello
Missoula
Great Falls
Havre
Billings
Lander
Williston
Bismarck
Rapid City
Cheyenne
Denver
Grand Junction
Albuquerque
El Paso
North Platte
Dodge City
Amarillo
Oklahoma City
Abilene
Fort Worth
Del Rio
San Antonio
Houston
Brownsville
Fargo
Huron
Minneapolis
International Falls
Duluth
Lincoln
Kansas City
Des Moines
Dubuque
Little Rock
Shreveport
Green Bay
Chicago
Springfield
St. Louis
Memphis
Jackson
New Orleans
Ste. Marie
Lansing
Detroit
Indianapolis
Nashville
Birmingham
Pensacola
Columbus
Lexington
Knoxville
Atlanta
Charleston
Pittsburgh
Buffalo
Harrisburg
Washington, D.C.
Richmond
Norfolk
Raleigh
Wilmington
Columbia
Jacksonville
Tampa
Miami
Philadelphia
New York
New Haven
Hartford
Albany
Boston
Burlington
Concord
Portland
Caribou

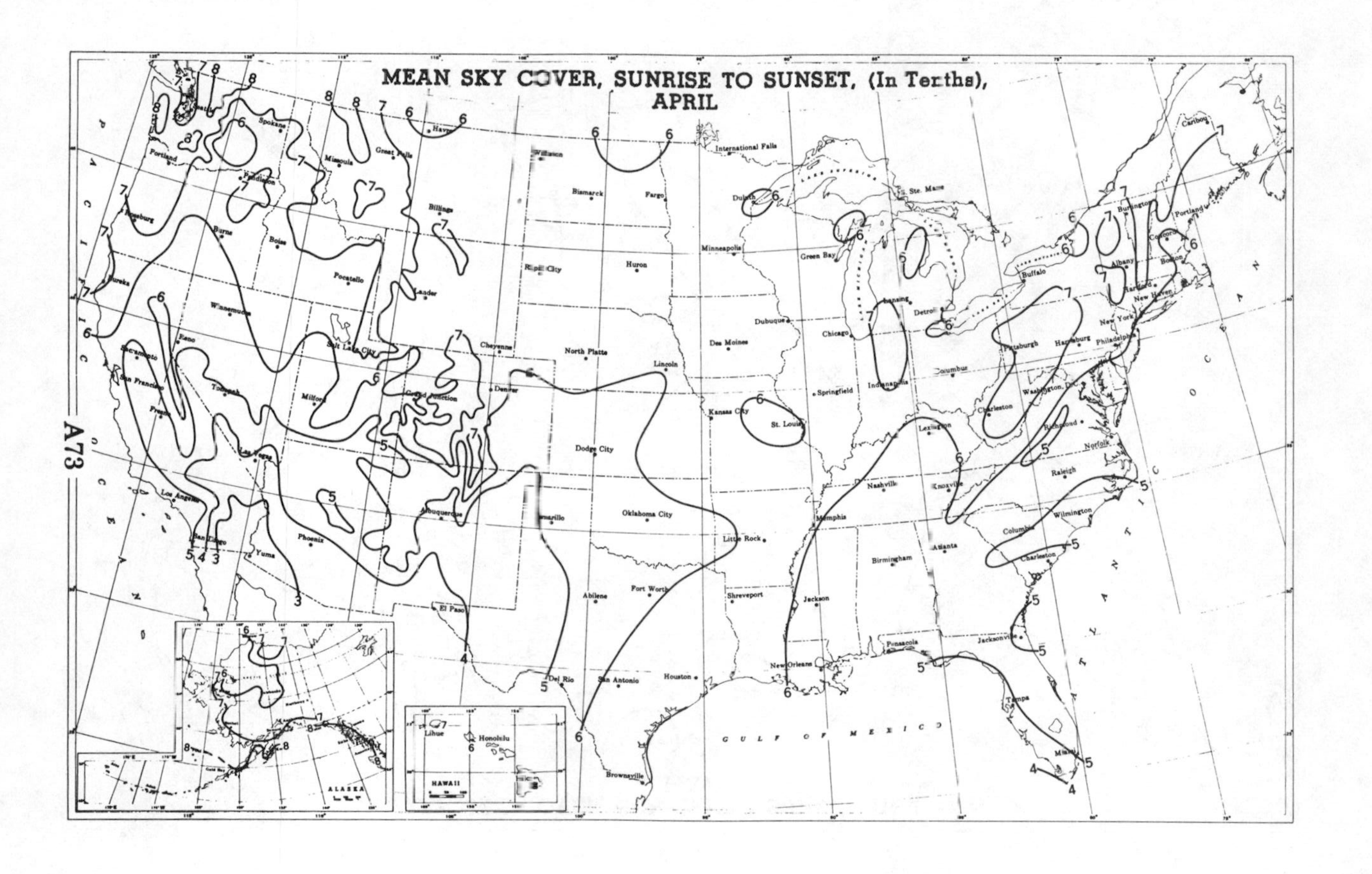
MEAN SKY COVER, SUNRISE TO SUNSET, (In Tenths),
APRIL
PACIFIC OCEAN
ATLANTIC OCEAN
GULF OF MEXICO
Portland
Roseburg
Eureka
Sacramento
San Francisco
Fresno
Los Angeles
San Diego
Reno
Tonopah
Las Vegas
Yuma
Phoenix
Spokane
Pendleton
Burns
Boise
Winnemucca
Missoula
Great Falls
Havre
Billings
Pocatello
Lander
Salt Lake City
Milford
Grand Junction
Cheyenne
Denver
Albuquerque
El Paso
Williston
Bismarck
Fargo
Rapid City
Huron
North Platte
Lincoln
Dodge City
Amarillo
Oklahoma City
Abilene
Fort Worth
Del Rio
San Antonio
Houston
Brownsville
International Falls
Duluth
Minneapolis
Des Moines
Dubuque
Kansas City
St. Louis
Springfield
Little Rock
Shreveport
Memphis
Jackson
New Orleans
Green Bay
Chicago
Lansing
Detroit
Indianapolis
Columbus
Nashville
Knoxville
Lexington
Birmingham
Atlanta
Pensacola
Jacksonville
Tampa
Miami
Sault Ste. Marie
Buffalo
Pittsburgh
Harrisburg
Charleston
Washington, D.C.
Richmond
Norfolk
Raleigh
Wilmington
Columbia
Charleston
Philadelphia
New York
New Haven
Hartford
Albany
Burlington
Concord
Boston
Portland
Caribou
ALASKA
HAWAII
Lihue
Honolulu

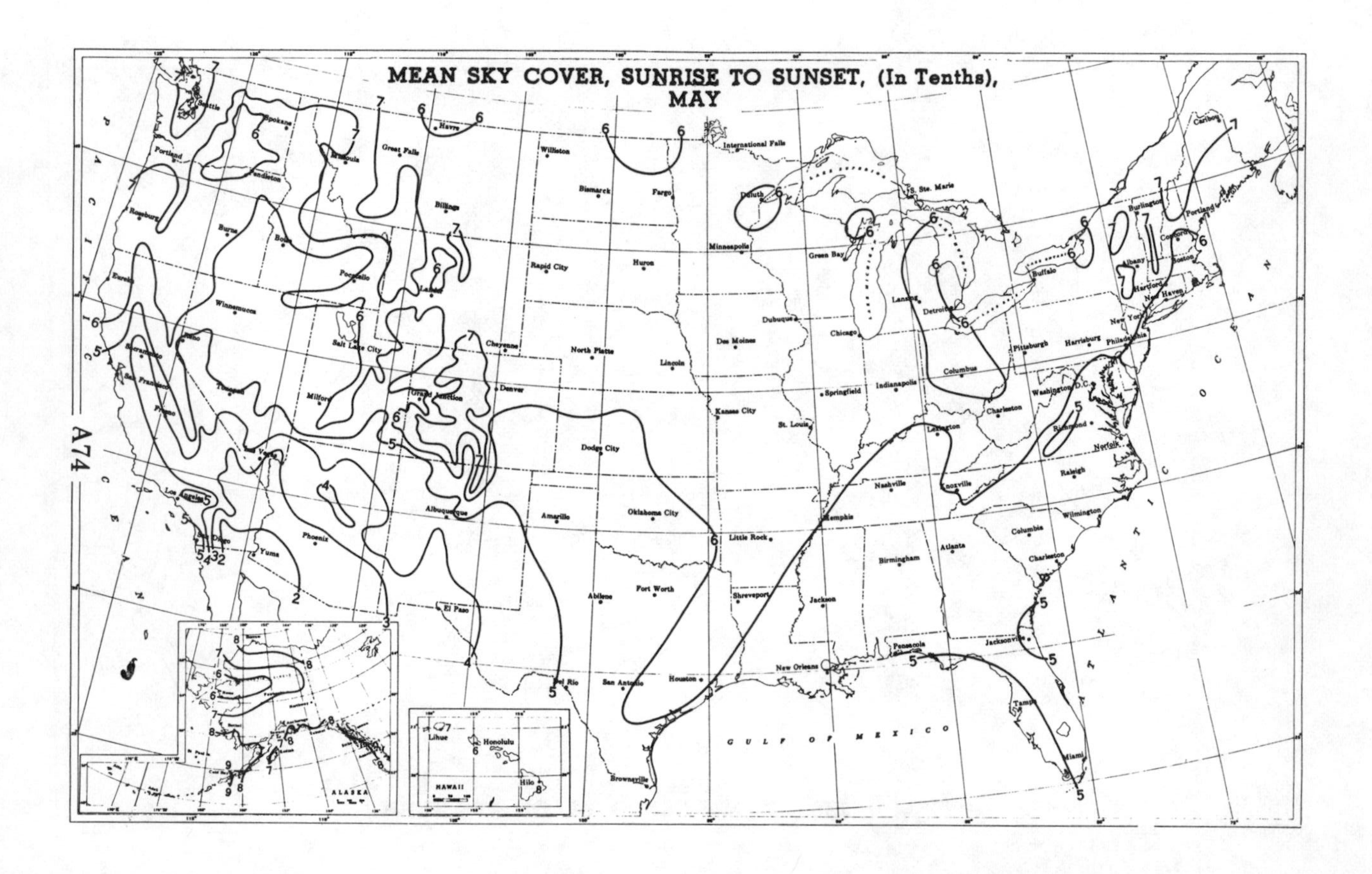

MEAN SKY COVER, SUNRISE TO SUNSET, (In Tenths), MAY

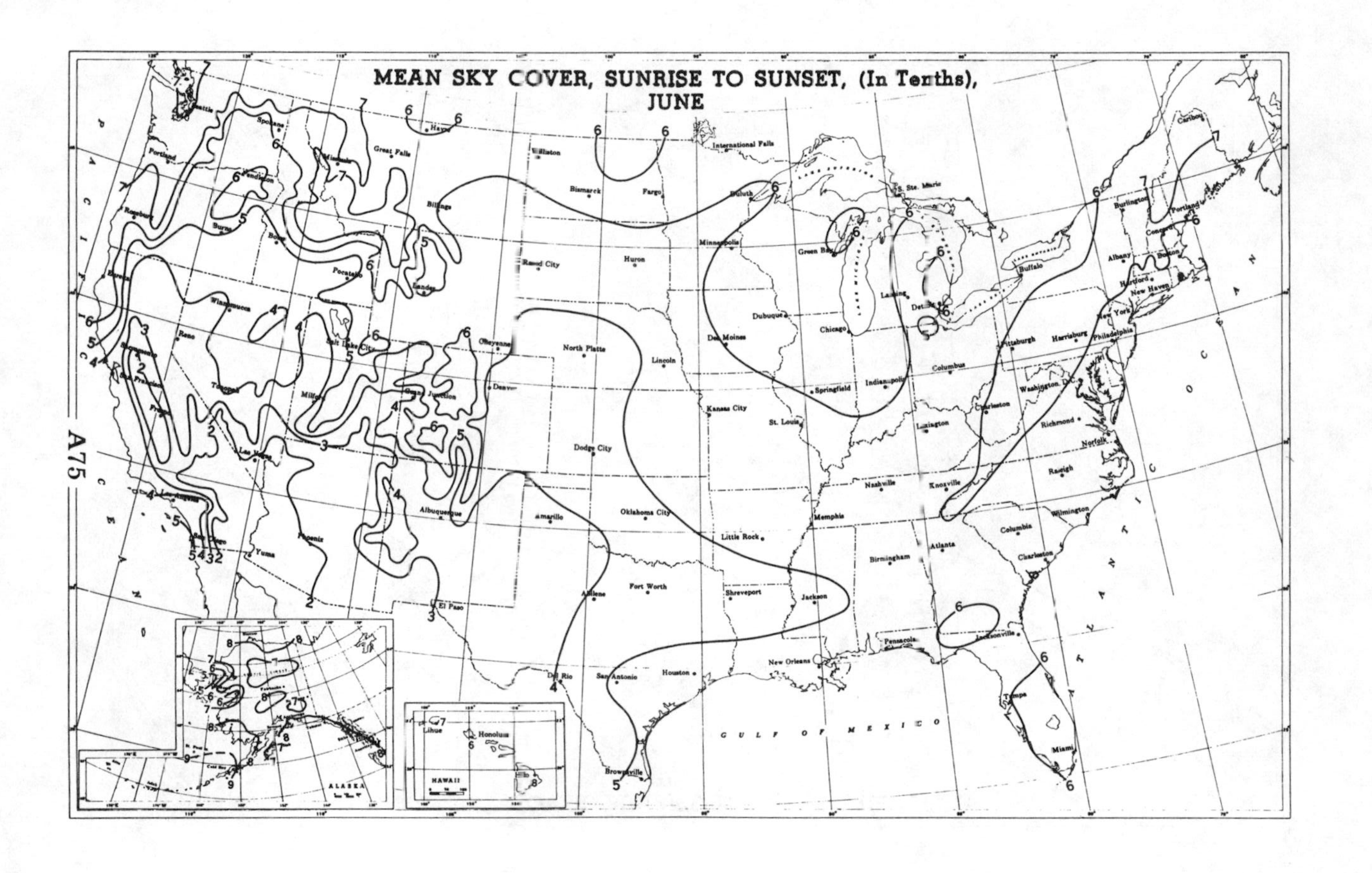
MEAN SKY COVER, SUNRISE TO SUNSET, (In Tenths),
JUNE
Seattle
Portland
Spokane
Pendleton
Roseburg
Burns
Boise
Missoula
Great Falls
Havre
Billings
Pocatello
Lander
Winnemucca
Reno
Sacramento
San Francisco
Fresno
Tonopah
Salt Lake City
Milford
Las Vegas
Los Angeles
San Diego
Yuma
Phoenix
Grand Junction
Cheyenne
Denver
Albuquerque
El Paso
Williston
Bismarck
Fargo
Rapid City
Huron
North Platte
Lincoln
Dodge City
Amarillo
Oklahoma City
Abilene
Fort Worth
Del Rio
San Antonio
Houston
Brownsville
International Falls
Duluth
Minneapolis
Des Moines
Dubuque
Kansas City
St. Louis
Springfield
Chicago
Green Bay
Sault Ste. Marie
Lansing
Detroit
Indianapolis
Columbus
Lexington
Nashville
Memphis
Little Rock
Shreveport
Jackson
New Orleans
Birmingham
Atlanta
Pensacola
Jacksonville
Tampa
Miami
Knoxville
Charleston
Columbia
Wilmington
Raleigh
Richmond
Norfolk
Washington, D.C.
Pittsburgh
Harrisburg
Philadelphia
New York
Buffalo
Albany
Hartford
New Haven
Boston
Concord
Portland
Burlington
Caribou
GULF OF MEXICO
ATLANTIC OCEAN
PACIFIC OCEAN
ALASKA
HAWAII
Lihue
Honolulu

MEAN SKY COVER, SUNRISE TO SUNSET, (In Tenths), JULY

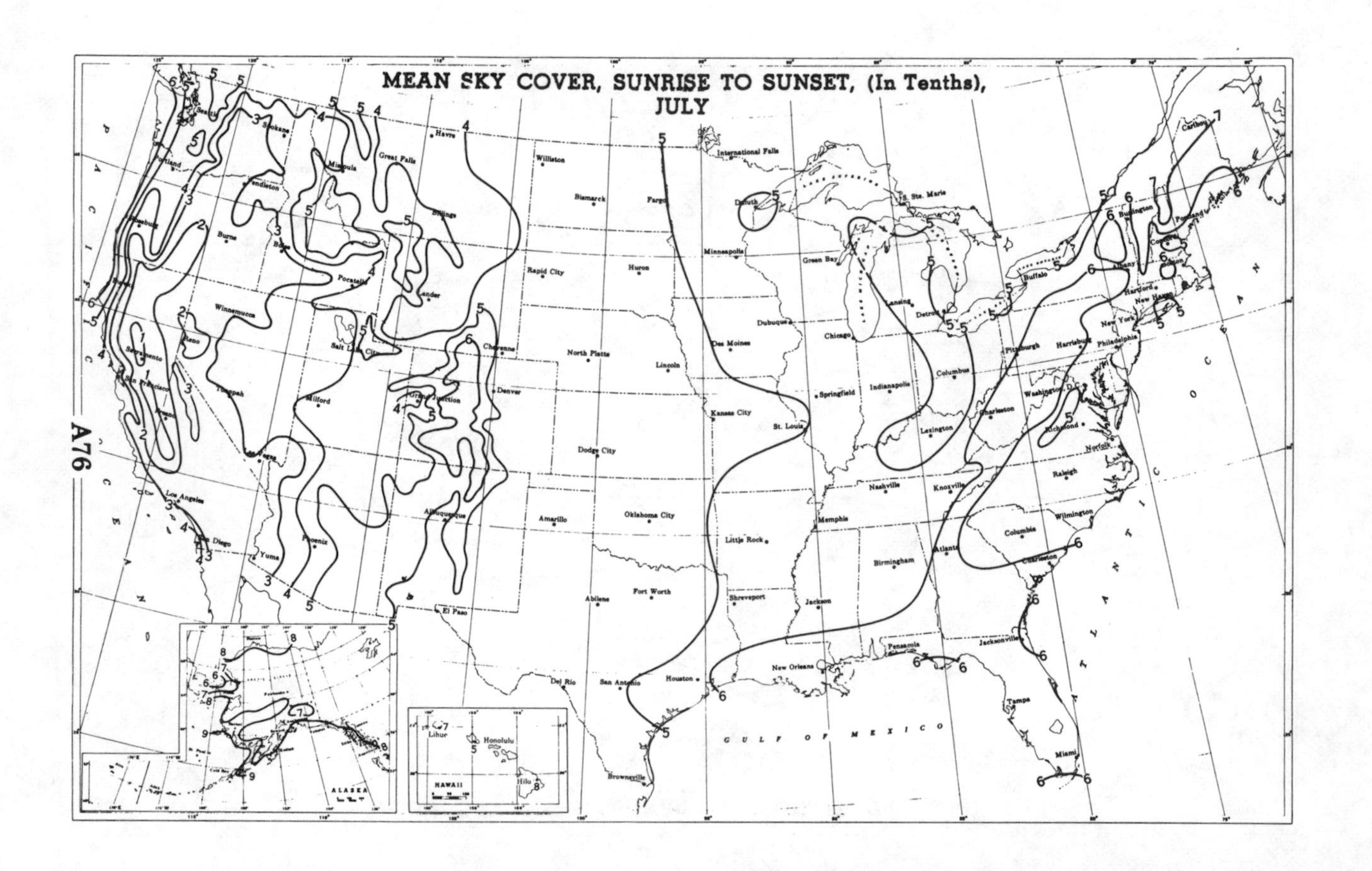

A76

MEAN SKY COVER, SUNRISE TO SUNSET, (In Tenths), AUGUST

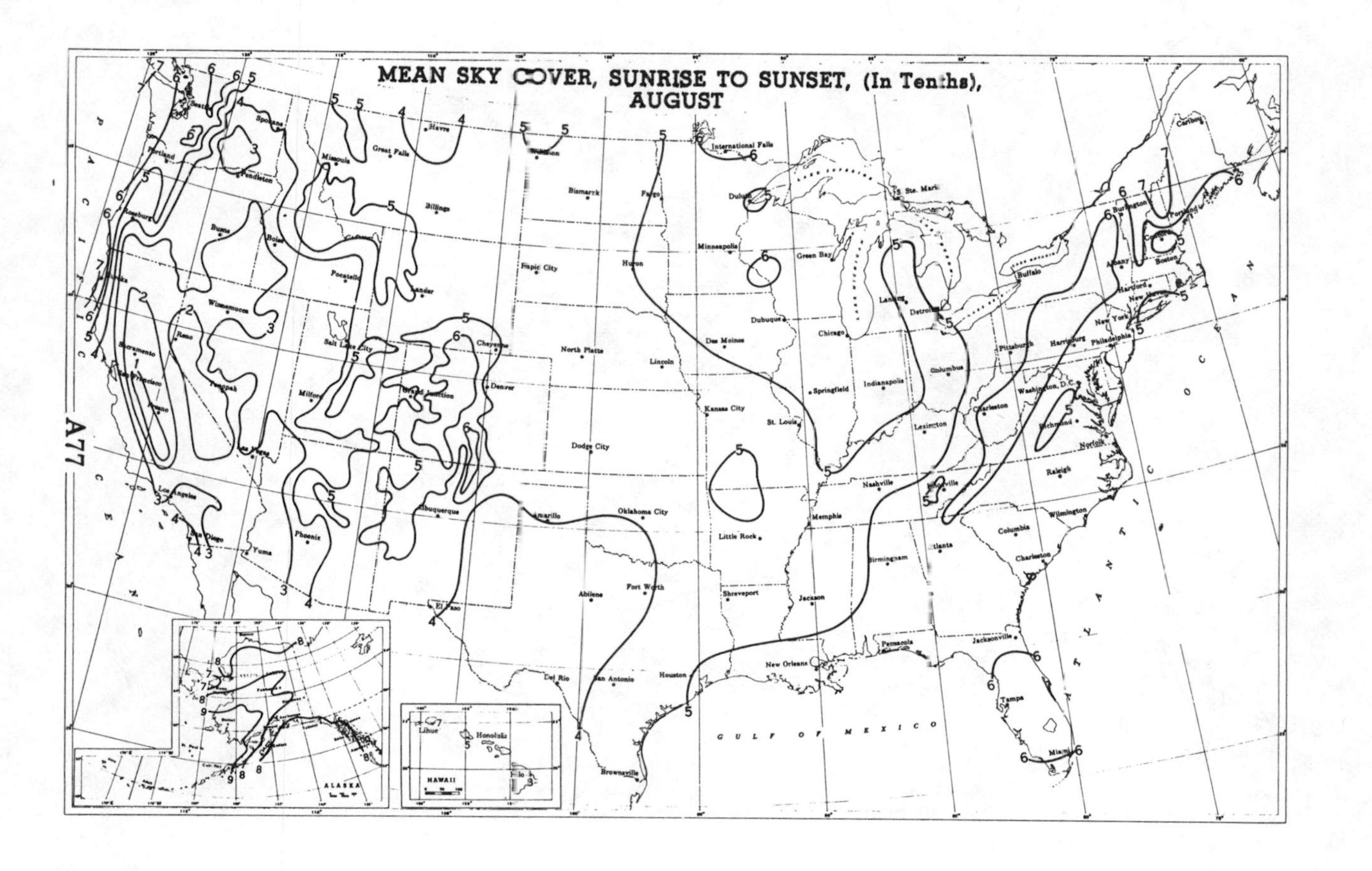

MEAN SKY COVER, SUNRISE TO SUNSET, (In Tenths), SEPTEMBER

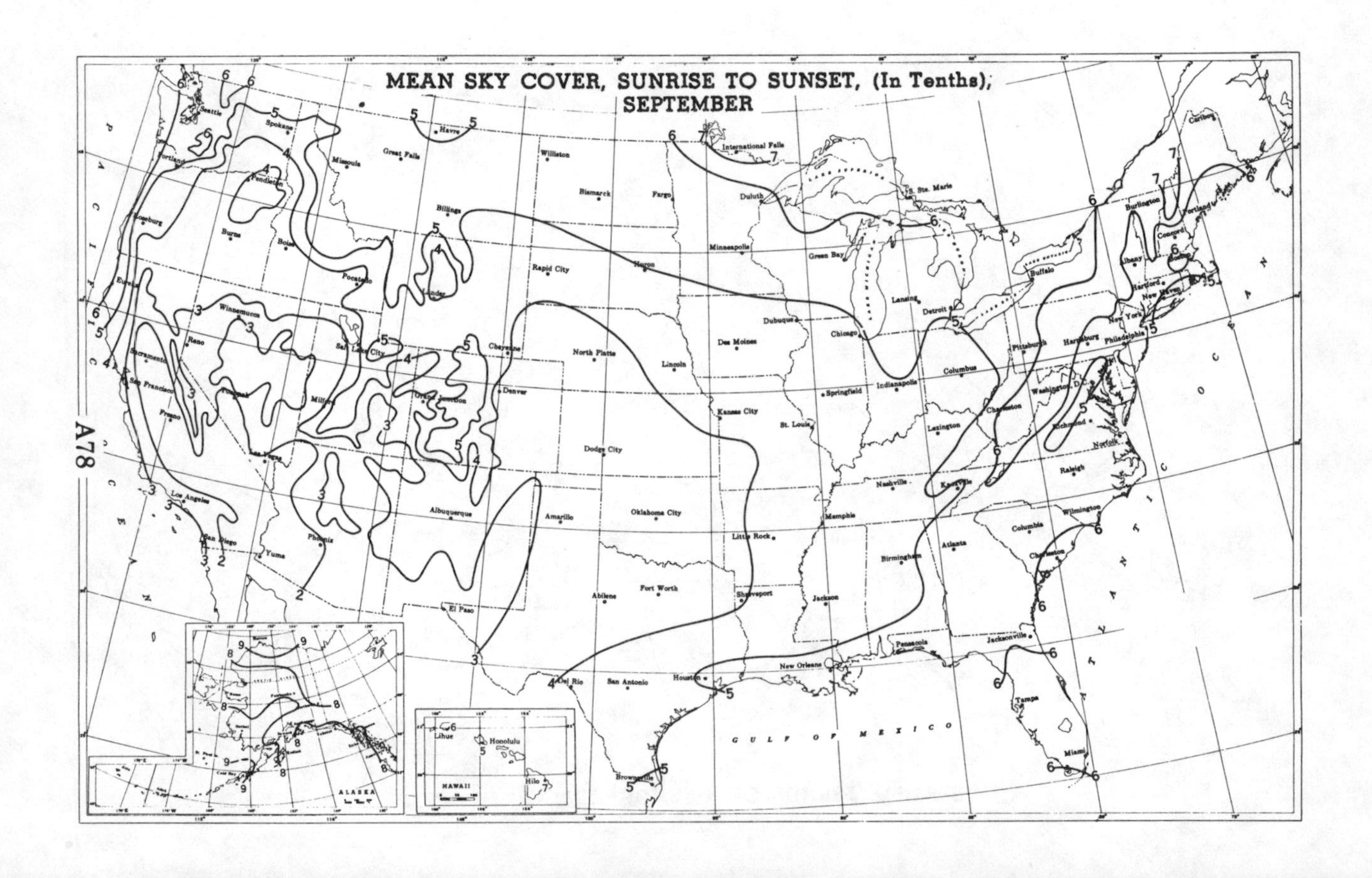

MEAN SKY COVER, SUNRISE TO SUNSET, (In Tenths), OCTOBER

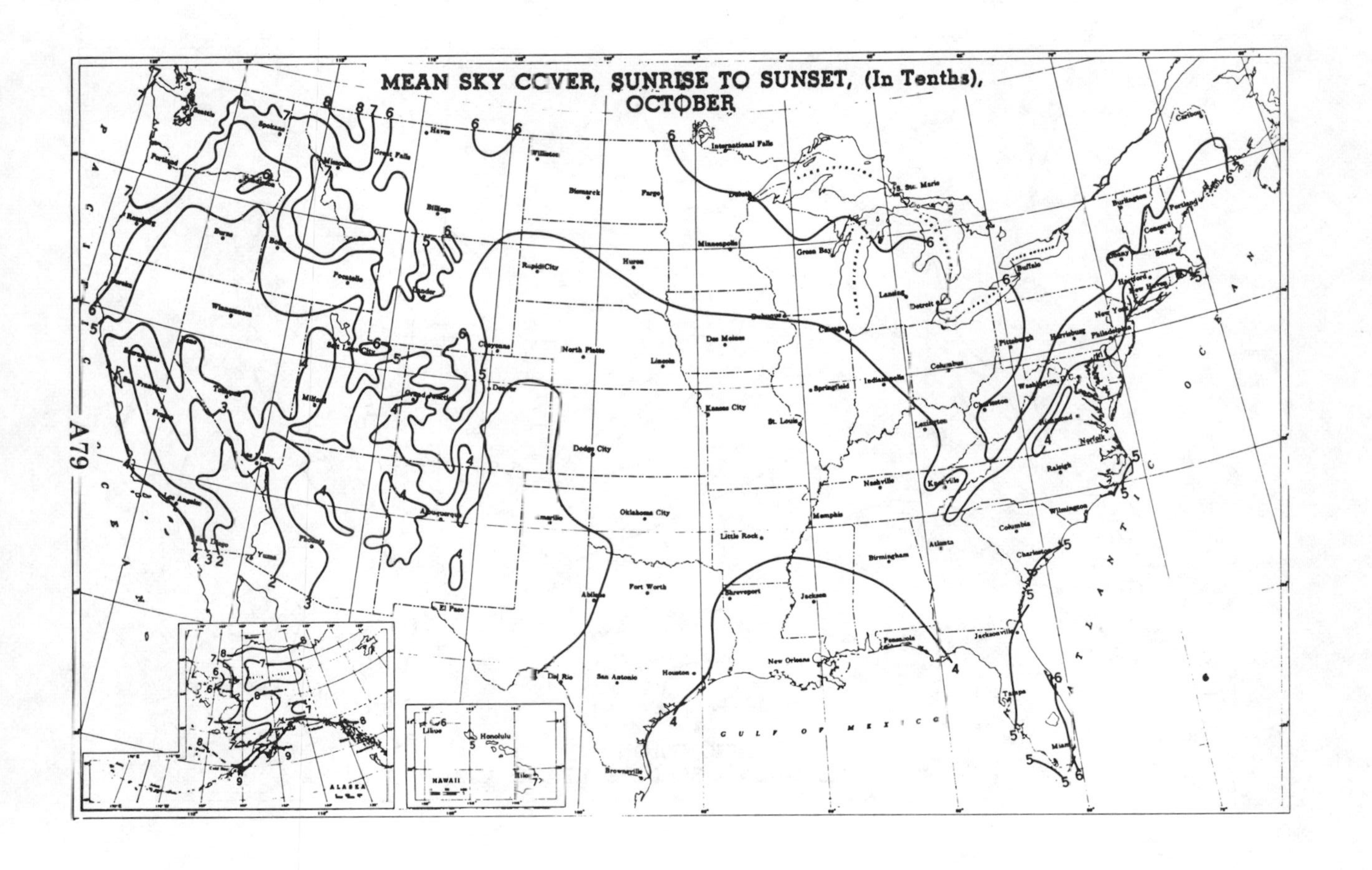

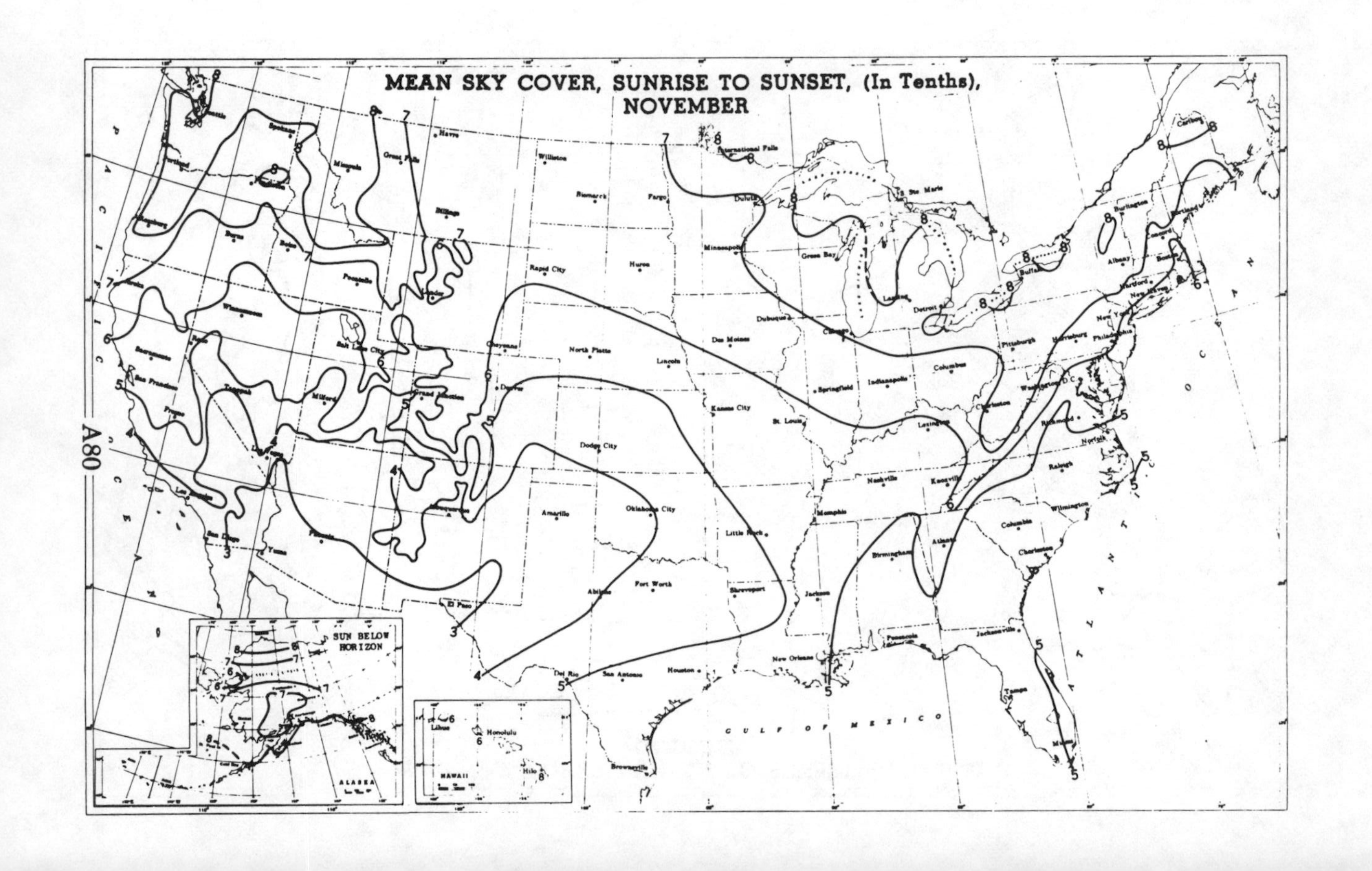

MEAN SKY COVER, SUNRISE TO SUNSET, (In Tenths),
NOVEMBER
PACIFIC OCEAN
ATLANTIC OCEAN
GULF OF MEXICO
SUN BELOW HORIZON
ALASKA
HAWAII
Honolulu
Hilo
Lihue
Seattle
Portland
Spokane
Missoula
Great Falls
Havre
Billings
Williston
Bismarck
Fargo
International Falls
Duluth
Sault Ste Marie
Burlington
Caribou
Portland
Concord
Boston
Albany
Hartford
New Haven
New York
Philadelphia
Harrisburg
Pittsburgh
Buffalo
Detroit
Lansing
Green Bay
Minneapolis
Huron
Rapid City
Pocatello
Boise
Salt Lake City
Winnemucca
Reno
Sacramento
San Francisco
Fresno
Tonopah
Milford
Grand Junction
Cheyenne
Denver
North Platte
Lincoln
Des Moines
Dubuque
Chicago
Springfield
Indianapolis
Columbus
Washington D.C.
Charleston
Richmond
Norfolk
Raleigh
Lexington
St. Louis
Kansas City
Dodge City
Las Vegas
Los Angeles
San Diego
Yuma
Phoenix
Albuquerque
Amarillo
Oklahoma City
Little Rock
Memphis
Nashville
Knoxville
Wilmington
Columbia
Charleston
Atlanta
Birmingham
Jackson
Shreveport
Fort Worth
Abilene
El Paso
Del Rio
San Antonio
Houston
New Orleans
Pensacola
Jacksonville
Tampa
Miami
Brownsville

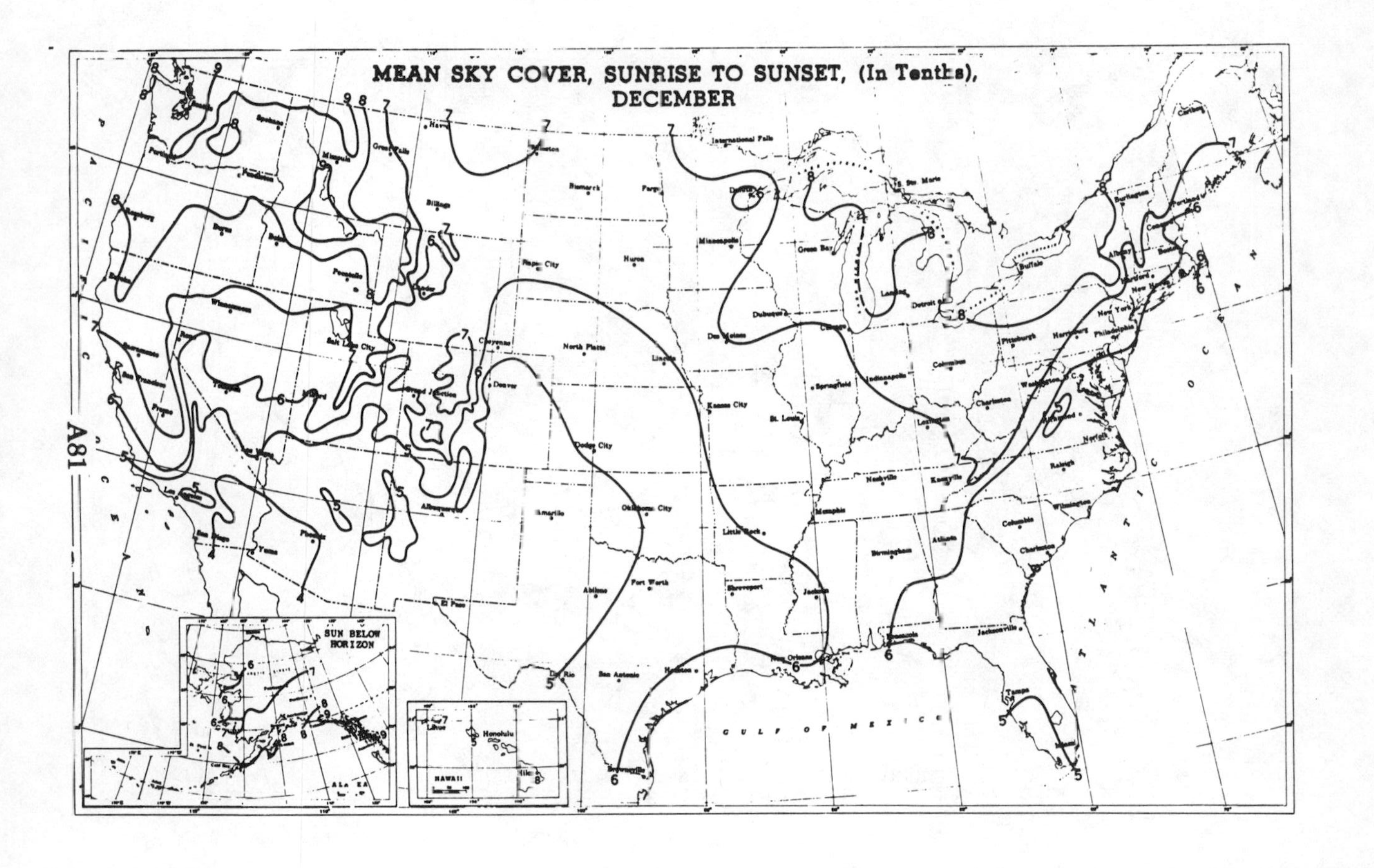
MEAN SKY COVER, SUNRISE TO SUNSET, (In Tenths),
DECEMBER
GULF OF MEXICO
PACIFIC OCEAN
ATLANTIC OCEAN
SUN BELOW HORIZON
HAWAII

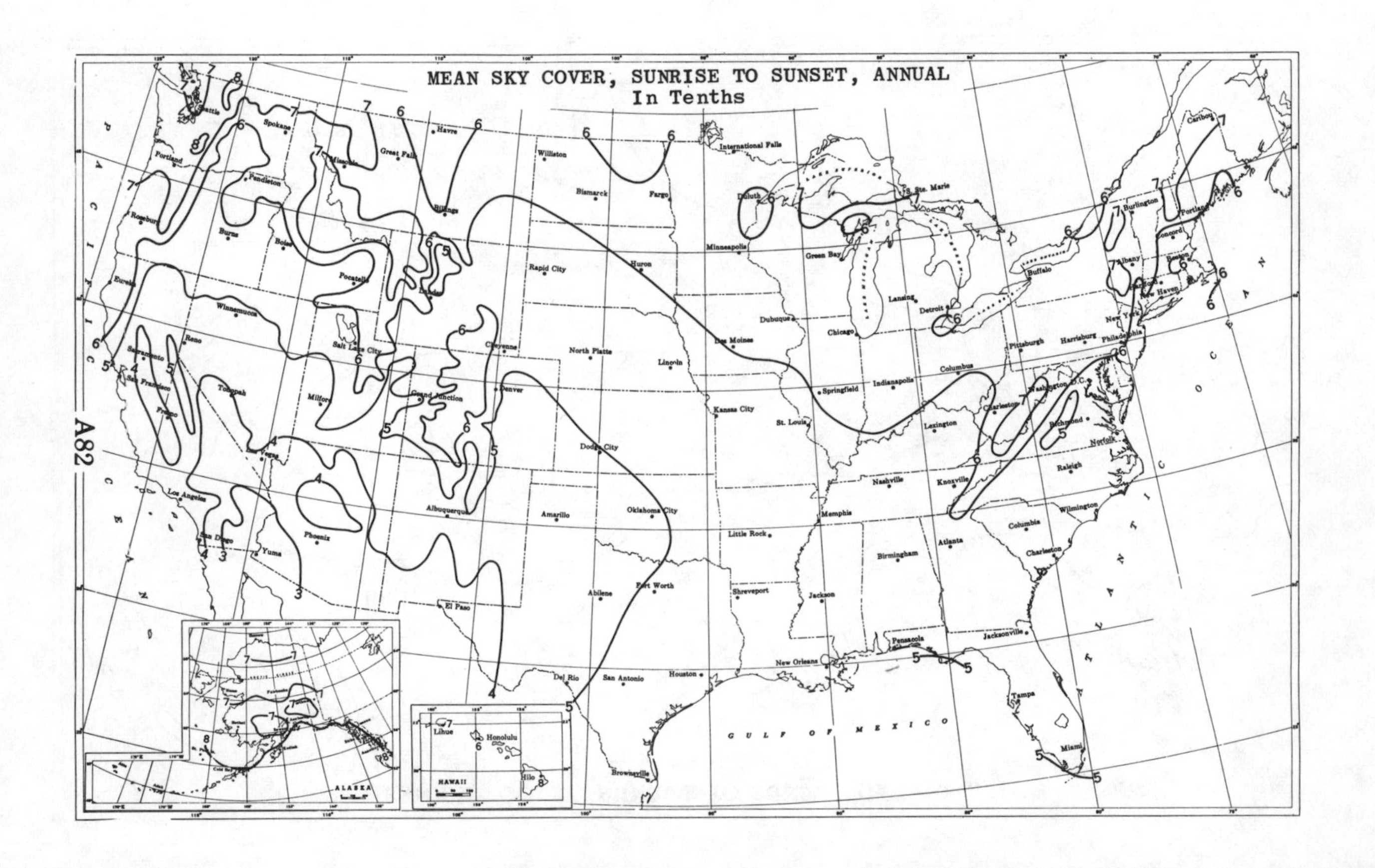
MEAN SKY COVER, SUNRISE TO SUNSET, ANNUAL
In Tenths
PACIFIC OCEAN
ATLANTIC OCEAN
GULF OF MEXICO
Seattle
Spokane
Havre
Great Falls
Williston
International Falls
Caribou
Portland
Pendleton
Missoula
Billings
Bismarck
Fargo
Duluth
Sault Ste. Marie
Burlington
Roseburg
Burns
Boise
Minneapolis
Green Bay
Albany
Concord
Boston
Eureka
Pocatello
Rapid City
Huron
Buffalo
Hartford
New Haven
Winnemucca
Lander
Dubuque
Lansing
Detroit
New York
Reno
Sacramento
Salt Lake City
Cheyenne
North Platte
Lincoln
Des Moines
Chicago
Pittsburgh
Harrisburg
Philadelphia
San Francisco
Tonopah
Fresno
Milford
Grand Junction
Denver
Kansas City
Springfield
Indianapolis
Columbus
Washington, D.C.
Charleston
Richmond
Las Vegas
Dodge City
St. Louis
Lexington
Norfolk
Los Angeles
Albuquerque
Amarillo
Oklahoma City
Nashville
Knoxville
Raleigh
Memphis
Wilmington
Columbia
Little Rock
San Diego
Yuma
Phoenix
Atlanta
Birmingham
Charleston
Abilene
Fort Worth
Shreveport
Jackson
El Paso
Jacksonville
Pensacola
New Orleans
Del Rio
San Antonio
Houston
Tampa
Miami
Brownsville
ALASKA
HAWAII
Lihue
Honolulu
Hilo

MEAN SKY COVER, SUNRISE TO SUNSET

STATE AND STATION	YEARS	JAN.	FEB.	MAR.	APR.	MAY	JUNE	JULY	AUG.	SEPT.	OCT.	NOV.	DEC.	ANNUAL
ALA. BIRMINGHAM	54	6.2	5.9	5.6	5.2	5.2	5.2	5.7	5.3	4.8	4.1	4.7	6.1	5.3
MOBILE	11	6.5	6.6	6.0	5.7	5.6	5.7	6.8	5.4	5.8	4.3	4.9	6.1	5.8
MONTGOMERY	83	5.9	5.7	5.2	4.8	4.7	5.1	5.5	5.2	4.5	3.9	4.5	5.7	5.1
ALASKA. ANCHORAGE	19	6.8	6.8	6.6	6.9	7.6	7.5	7.6	7.8	7.8	7.6	7.3	7.4	7.3
ANNETTE	12	7.7	8.0	7.7	7.6	7.9	8.0	8.0	7.8	7.7	8.6	8.5	8.6	8.0
BARROW	18	*	5.3	5.0	6.0	8.5	8.0	8.2	8.9	9.2	8.8	*	*	-
BARTER ISLAND	9	*	4.7	5.8	6.5	8.5	8.1	7.9	8.3	8.6	8.5	*	*	-
BETHEL	16	6.3	6.2	6.5	6.9	7.9	8.1	8.5	9.0	8.2	7.7	7.2	6.7	7.4
CORDOVA	14	7.2	7.1	7.1	7.4	8.5	8.3	8.6	8.1	8.2	8.0	7.9	7.8	7.9
FAIRBANKS	16	6.5	6.3	6.2	6.2	7.0	7.3	7.3	7.8	7.9	7.9	7.1	7.1	7.1
JUNEAU	16	7.9	8.0	7.8	8.2	7.8	7.8	8.3	8.0	8.4	8.6	8.5	8.5	8.2
KOTZEBUE	16	5.6	5.5	6.0	5.8	6.4	6.9	7.7	8.3	7.5	7.2	6.6	5.9	6.6
MCGRATH	16	6.6	6.3	6.5	6.1	7.5	7.8	7.8	8.5	8.1	8.4	7.4	6.8	7.3
NOME	25	5.9	5.8	5.9	6.5	6.7	6.6	7.9	8.1	7.5	7.1	6.5	6.6	6.8
ST. PAUL ISLAND	30	8.1	8.0	7.7	8.1	8.7	8.9	9.1	9.1	8.6	8.5	8.3	8.4	8.5
ARIZ. FLAGSTAFF	10	5.9	4.9	5.0	4.9	4.1	2.4	5.4	5.3	2.9	3.1	3.4	4.4	4.3
PHOENIX	14	5.1	4.5	4.3	3.8	2.9	2.0	3.8	3.5	2.0	2.8	3.1	3.6	3.5
TUCSON	16	5.0	4.5	4.6	3.7	2.8	2.2	5.5	4.7	2.7	3.1	3.0	4.4	3.9
YUMA	80	2.4	2.5	2.2	1.6	1.1	0.6	1.7	1.8	1.1	1.3	1.7	2.5	1.7
ARK. FT. SMITH	67	5.4	5.5	5.4	5.4	5.2	4.6	4.4	4.2	3.9	3.3	4.5	5.3	4.8
LITTLE ROCK	55	6.2	6.3	6.5	6.4	6.2	5.7	6.0	5.4	4.8	3.9	5.0	5.9	5.7
CALIF. BAKERSFIELD	14	6.2	5.5	5.4	4.4	3.2	1.3	1.1	1.1	1.3	2.7	4.1	6.3	3.6
BISHOP	12	5.3	4.8	4.7	4.7	4.1	2.3	2.4	1.9	2.0	2.8	3.5	4.9	3.6
EUREKA	17	7.3	7.3	7.3	7.2	6.8	6.4	6.6	6.9	6.0	6.5	7.0	7.6	6.9
FRESNO	22	6.7	6.1	5.3	4.4	3.4	1.9	1.1	1.1	1.5	2.8	4.4	7.0	3.8
LOS ANGELES	19	4.6	4.8	4.8	5.3	4.8	4.1	2.8	2.7	2.8	3.9	3.4	4.5	4.0
RED BLUFF	15	6.6	6.3	6.1	5.4	4.6	3.2	1.3	1.6	2.2	4.0	5.5	6.9	4.5
SACRAMENTO	50	6.0	5.2	4.4	3.5	2.7	1.6	0.7	0.7	1.4	2.4	4.0	5.9	3.2
SAN DIEGO	19	5.1	5.0	4.9	5.7	5.5	5.1	4.4	4.3	3.7	4.3	3.7	4.7	4.7
SAN FRANCISCO	49	5.7	5.5	5.0	4.5	4.4	3.7	4.2	4.6	3.8	4.0	4.5	5.6	4.6
COLO. COLORADO SPRINGS	10	4.9	5.1	5.7	6.0	6.0	4.5	4.9	4.7	3.5	3.7	4.7	4.5	4.9
DENVER	48	4.5	4.8	5.3	5.7	5.7	4.7	4.6	4.7	4.0	4.1	4.2	4.4	4.7
GRAND JUNCTION	57	5.3	5.4	5.3	5.4	4.8	3.6	3.9	4.0	3.3	3.5	4.2	5.1	4.5
CONN. HARTFORD	51	6.2	5.7	5.8	6.1	6.0	5.8	5.7	5.6	5.4	5.3	6.2	6.2	5.8
D.C. WASHINGTON	70	6.1	5.6	5.6	5.4	5.4	5.1	5.1	5.1	4.0	4.7	5.1	5.0	5.0
FLA. APALACHICOLA	26	5.2	5.4	5.5	4.8	4.3	5.2	6.0	5.7	5.6	3.9	4.4	5.7	5.1
JACKSONVILLE	68	5.3	5.2	4.9	4.6	4.7	5.6	5.9	5.6	5.7	5.1	4.7	5.4	5.2
KEY WEST	58	4.3	3.8	3.7	3.8	4.4	5.4	5.3	5.3	5.5	4.9	4.2	4.3	4.6
MIAMI	48	5.0	4.6	4.7	4.9	5.4	6.2	6.1	5.8	6.2	5.8	5.2	5.2	5.4
TAMPA	34	5.3	4.7	4.8	4.7	4.6	5.8	6.3	6.1	6.0	4.7	4.6	5.2	5.2
GA. ATLANTA	25	6.3	6.2	6.1	5.5	5.5	5.8	6.3	5.7	5.3	4.5	5.1	6.2	5.7
HAWAII. HILO	13	6.8	7.0	7.8	8.3	8.2	7.6	7.8	7.9	7.2	7.4	7.6	7.6	7.6
HONOLULU	13	5.3	6.0	5.9	6.0	6.0	5.5	5.0	5.1	4.8	5.4	5.5	5.4	5.5
LIHUE	10	6.2	6.5	6.5	6.9	7.0	6.6	6.5	6.6	5.7	6.2	6.4	6.6	6.5
IDAHO. BOISE	20	7.5	7.2	6.8	6.2	5.8	4.7	2.7	3.1	3.5	5.0	6.6	7.6	5.6
POCATELLO	21	7.7	7.5	6.8	6.3	6.0	4.8	3.4	3.4	3.8	5.1	6.5	7.7	5.8
ILL. CAIRO	66	6.5	6.3	6.3	6.0	5.9	5.6	5.2	5.0	4.6	4.4	5.5	6.3	5.6
CHICAGO	84	6.4	6.2	6.1	5.8	5.3	5.2	4.3	4.6	4.7	5.0	6.3	6.6	5.5
SPRINGFIELD	31	6.4	6.4	6.4	6.1	5.9	5.7	4.7	4.9	4.9	4.8	6.1	6.6	5.7
IND. FT. WAYNE	44	7.1	7.0	6.6	6.4	5.8	5.6	4.7	5.0	5.0	5.4	6.7	7.4	6.1
INDIANAPOLIS	51	7.1	6.8	6.6	6.4	6.0	5.6	5.0	5.1	4.8	5.0	6.3	7.1	6.0
IOWA. DES MOINES	73	5.4	5.5	5.8	5.7	5.6	5.3	4.3	4.4	4.5	4.4	5.3	5.8	5.2
DUBUQUE	62	6.1	6.0	6.2	5.9	5.9	5.6	4.7	5.0	5.1	5.2	6.3	6.5	5.7
SIOUX CITY	66	5.6	5.8	6.1	5.9	5.9	5.3	4.3	4.6	4.5	4.6	5.7	5.9	5.4
KANS. CONCORDIA	16	5.5	5.8	5.9	6.0	6.0	5.0	4.5	4.2	4.0	4.0	4.8	5.3	5.1
DODGE CITY	67	4.0	4.5	4.7	4.8	4.9	4.3	3.9	3.7	3.5	3.4	3.6	3.9	4.1
GOODLAND	11	5.8	5.9	6.0	6.0	6.2	4.4	4.4	4.4	3.6	3.9	4.8	5.1	5.0
WICHITA	66	5.0	5.1	5.2	5.3	5.3	4.6	3.9	3.8	3.9	4.0	4.4	5.0	4.6
KY. LOUISVILLE	54	6.8	6.3	6.1	6.0	5.5	5.3	4.8	4.8	4.5	4.6	5.7	6.7	5.6
LA. NEW ORLEANS	44	6.0	6.0	5.8	5.3	5.1	5.4	6.1	5.8	5.2	4.0	4.7	5.9	5.4

MEAN SKY COVER, SUNRISE TO SUNSET

STATE AND STATION	YEARS	JAN.	FEB.	MAR.	APR.	MAY	JUNE	JULY	AUG.	SEPT.	OCT.	NOV.	DEC.	ANNUAL
SHREVEPORT	83	5.8	5.7	5.4	5.1	4.8	4.4	4.5	4.0	4.0	3.7	4.7	5.6	4.8
MAINE. CARIBOU	14	7.2	6.8	6.8	7.3	7.4	7.5	7.0	6.8	6.6	6.8	8.1	7.3	7.1
EASTPORT	61	6.9	6.5	6.6	6.7	6.9	6.9	6.5	6.3	6.3	6.6	7.6	7.2	6.8
MASS. BOSTON	64	6.0	5.6	5.6	5.8	5.8	5.7	5.6	5.3	5.1	5.3	6.0	6.0	5.7
MICH. ALPENA	45	7.7	6.9	6.4	6.0	5.8	5.5	4.8	5.1	5.8	6.3	7.9	8.0	6.4
DETROIT	63	7.4	6.9	6.5	6.1	5.7	5.2	4.7	4.8	5.0	5.4	7.1	7.6	6.0
GRAND RAPIDS	51	8.2	7.5	6.8	6.3	5.9	5.5	4.7	5.0	5.4	6.0	7.5	8.3	6.4
MARQUETTE	66	7.7	7.3	6.9	6.3	6.3	5.9	5.5	5.8	6.3	7.0	8.0	8.0	6.8
SAULT STE. MARIE	64	7.9	6.8	6.3	5.9	6.0	5.5	5.2	5.6	6.3	7.2	8.4	8.3	6.6
MINN. DULUTH	64	6.1	5.5	5.6	5.5	5.8	5.6	4.8	5.0	5.4	5.9	6.8	6.1	5.7
MINNEAPOLIS	24	6.5	6.2	6.7	6.5	6.4	6.0	4.9	5.1	5.1	5.4	6.9	6.9	6.1
MISS. MERIDIAN	22	6.1	6.0	5.5	5.1	4.9	4.9	5.6	4.9	4.5	3.9	4.6	6.1	5.2
VICKSBURG	55	5.7	5.8	5.6	5.3	5.1	4.8	5.5	4.9	4.1	3.4	4.5	5.6	5.0
MO. KANSAS CITY	66	5.4	5.4	5.5	5.5	5.3	4.9	4.1	4.1	4.2	4.1	5.0	5.5	4.9
ST. LOUIS	24	6.3	5.9	6.0	5.7	5.3	5.0	4.2	4.5	4.1	4.1	5.4	6.4	5.2
SPRINGFIELD	50	5.7	5.3	5.2	5.0	4.6	4.1	3.6	3.9	3.8	4.0	4.8	5.6	4.6
MONT. HAVRE	56	6.3	5.9	5.9	5.7	5.5	5.3	3.6	3.8	4.8	5.4	6.2	6.3	5.4
HELENA	65	6.8	6.6	6.6	6.5	6.5	6.0	4.0	4.2	5.1	5.7	6.6	6.8	6.0
KALISPELL	50	7.7	6.7	6.2	5.7	5.6	5.3	3.5	3.9	4.9	5.7	7.6	8.0	5.9
MILES CITY	10	6.8	6.9	7.0	6.8	6.3	5.7	3.9	4.1	5.2	5.3	6.6	6.5	5.9
NEBR. LINCOLN	60	5.5	5.7	5.8	5.9	5.9	5.2	4.3	4.6	4.4	4.4	5.3	5.6	5.2
NORTH PLATTE	11	6.2	6.2	6.4	6.6	6.6	5.3	4.9	4.9	4.5	4.4	5.8	5.8	5.6
NEV. ELY	16	6.2	6.1	6.2	6.0	5.9	4.0	3.9	3.4	3.0	4.1	5.4	6.0	5.0
LAS VEGAS	11	5.2	4.2	4.1	3.8	3.3	1.6	2.8	2.3	1.5	2.3	3.2	4.4	3.2
RENO	35	5.3	5.3	4.9	4.6	4.1	3.1	1.8	1.8	2.1	3.4	4.3	5.5	3.9
WINNEMUCCA	24	6.4	6.3	6.0	5.8	5.1	4.0	2.3	2.2	2.4	4.0	5.3	6.6	4.7
N.H. CONCORD	41	5.6	5.0	4.9	5.2	5.2	4.8	4.7	4.8	5.0	5.3	6.0	5.9	5.2
N.J. ATLANTIC CITY	59	6.2	5.6	5.7	5.8	5.6	5.5	5.3	5.2	5.1	4.8	5.3	6.0	5.5
N.MEX. ALBUQUERQUE	25	4.3	4.5	4.6	4.5	4.3	3.4	4.4	4.5	3.2	2.9	3.1	4.0	4.0
RATON	8	5.1	5.4	5.2	5.9	6.1	4.5	5.3	5.4	3.9	4.0	4.4	4.2	5.0
ROSWELL	38	4.1	4.0	4.0	3.8	3.9	3.2	3.6	3.6	3.4	3.3	3.3	3.9	3.7
N.Y. ALBANY	65	6.6	6.0	5.9	5.8	5.6	5.3	5.1	5.0	5.0	5.4	6.4	6.7	5.7
BINGHAMTON	58	7.5	7.2	7.1	6.9	6.5	6.2	6.1	6.0	6.0	6.3	7.5	7.9	6.8
BUFFALO	57	8.1	7.5	7.0	6.5	6.0	5.5	5.2	5.4	5.6	6.2	7.7	8.2	6.6
CANTON	43	6.9	6.3	6.0	6.1	5.7	5.0	4.8	5.0	5.5	6.1	7.5	7.4	6.0
NEW YORK	69	6.2	5.8	5.9	5.2	5.8	5.7	5.6	5.2	5.0	5.0	5.8	6.0	5.7
SYRACUSE	55	7.9	7.5	7.1	6.6	6.0	5.7	5.4	5.6	5.8	6.3	7.7	8.1	6.6
N.C. ASHEVILLE	56	6.0	5.7	5.6	5.4	5.3	5.5	5.9	5.7	5.2	4.4	4.8	5.8	5.5
HATTERAS	24	6.1	5.9	5.9	5.1	5.3	5.4	5.8	5.6	5.7	5.1	5.1	5.9	5.6
RALEIGH	54	6.0	5.5	5.4	5.1	5.1	5.4	5.6	5.5	5.2	4.4	4.7	5.8	5.3
WILMINGTON	19	5.7	5.4	5.3	4.6	4.9	5.3	5.6	5.4	5.6	4.4	4.5	5.4	5.2
N.DAK. BISMARCK	15	6.7	6.6	6.9	6.6	6.4	6.2	4.5	4.8	5.4	5.6	6.6	6.6	6.1
DEVILS LAKE	55	6.1	6.0	6.1	5.8	5.8	5.6	4.5	4.7	5.3	5.6	6.6	6.3	5.7
FARGO	40	6.5	6.3	6.4	6.0	5.7	5.7	4.4	4.6	5.3	5.7	6.7	6.7	5.8
WILLISTON	18	6.8	6.9	7.1	6.6	6.4	6.6	4.7	5.0	5.9	6.0	6.9	6.9	6.3
OHIO. CINCINNATI	24	7.4	6.8	6.8	6.5	6.2	5.8	5.1	5.1	4.8	5.0	6.4	7.3	6.1
CLEVELAND	65	7.7	7.4	6.7	6.1	5.6	5.2	4.6	4.8	5.1	5.7	7.4	8.0	6.2
COLUMBUS	64	7.1	6.7	6.6	6.2	5.6	5.2	4.8	4.8	4.7	5.0	6.5	7.2	5.9
OKLA. OKLAHOMA CITY	23	5.5	5.7	5.4	5.5	5.6	4.7	4.2	4.3	4.1	4.1	4.6	5.2	4.9
OREG. BAKER	58	6.9	6.7	6.3	6.9	5.7	5.0	2.8	2.8	3.8	4.5	6.0	6.7	5.3
MEDFORD	29	8.2	7.6	7.2	6.6	5.8	4.9	2.1	2.2	3.3	5.6	7.4	8.6	5.8
PORTLAND	13	7.9	7.9	8.0	7.4	7.1	6.9	4.6	5.4	5.1	7.0	8.2	8.6	7.0
ROSEBURG	7	8.9	8.5	7.9	7.2	6.4	6.4	2.8	3.7	4.9	6.9	8.4	8.9	6.7
PA. HARRISBURG	58	6.6	6.1	6.1	6.2	6.0	5.9	5.6	5.6	5.1	5.2	6.2	6.5	5.9
PHILADELPHIA	63	6.1	5.8	5.8	5.8	5.8	5.8	5.7	5.6	5.2	4.9	5.6	6.0	5.7
PITTSBURGH	83	7.4	7.0	6.7	6.3	5.7	5.5	5.3	5.2	5.0	5.4	6.7	7.5	6.1
R.I. BLOCK ISLAND	27	6.0	5.5	5.5	5.7	5.7	5.3	5.2	5.0	4.8	4.6	5.7	5.9	5.4
PROVIDENCE	49	5.9	5.4	5.4	5.7	5.6	5.4	5.4	5.1	5.0	4.7	5.5	5.7	5.4
S.C. CHARLESTON	59	5.4	5.3	5.1	4.5	4.6	5.3	5.7	5.5	5.1	4.3	4.3	5.3	5.0
COLUMBIA	15	6.0	5.7	5.8	5.1	5.2	5.1	5.8	5.1	5.6	4.6	4.8	5.7	5.4

MEAN SKY COVER, SUNRISE TO SUNSET

STATE AND STATION	YEARS	JAN.	FEB.	MAR.	APR.	MAY	JUNE	JULY	AUG.	SEPT.	OCT.	NOV.	DEC.	ANNUAL
S.DAK. HURON	16	6.6	6.4	7.2	6.6	6.2	5.7	4.6	4.9	5.0	5.1	6.6	6.7	6.0
RAPID CITY	17	6.4	6.4	6.6	6.6	6.5	5.7	4.2	4.3	4.6	4.9	6.2	6.2	5.7
TENN. KNOXVILLE	78	5.9	5.8	5.5	5.2	4.9	5.0	5.1	5.0	4.1	3.7	4.7	5.7	5.0
MEMPHIS	14	7.2	6.5	6.5	6.0	6.0	5.2	5.3	4.7	4.5	4.1	5.2	6.2	5.6
NASHVILLE	36	6.8	6.5	6.1	5.8	5.6	5.2	5.2	5.0	4.7	4.3	5.5	6.5	5.6
TEX. ABILENE	25	5.6	5.6	5.4	4.9	5.2	4.2	4.2	4.2	4.4	3.8	4.3	5.1	4.7
AMARILLO	64	4.2	4.5	4.2	4.3	4.4	3.8	4.0	3.8	3.7	3.7	3.6	4.2	4.0
AUSTIN	29	6.2	6.1	5.7	5.6	5.5	4.7	4.5	4.0	4.5	4.2	5.1	5.8	5.2
BROWNSVILLE	32	6.5	6.2	6.1	5.9	5.4	4.7	4.5	4.3	5.0	4.4	5.7	6.4	5.4
CORPUS CHRISTI	23	6.5	6.7	6.3	6.2	5.8	4.6	4.5	4.2	4.7	3.9	5.2	6.4	5.4
DEL RIO	49	5.2	5.3	5.3	5.4	5.7	4.9	4.2	3.5	4.4	4.6	5.1	5.2	4.9
EL PASO	64	3.7	3.6	3.4	2.9	2.6	2.4	4.0	3.8	3.1	2.7	2.8	3.5	3.2
FT. WORTH	25	5.9	5.6	5.2	5.2	5.4	4.0	4.0	3.8	3.7	3.9	4.4	5.0	4.7
GALVESTON	52	5.9	5.9	5.7	5.4	4.8	4.1	4.7	4.5	4.3	3.6	4.7	5.9	5.0
HOUSTON	46	6.1	6.1	5.9	5.7	5.4	4.8	5.2	5.0	4.7	4.2	5.2	6.2	5.4
MIDLAND	11	5.4	5.3	5.1	4.9	4.7	3.8	4.6	3.9	3.5	3.6	3.5	4.3	4.4
SAN ANTONIO	17	6.4	6.5	6.3	6.4	6.2	5.4	5.0	4.6	4.9	4.6	5.4	5.7	5.6
UTAH. MODENA	46	5.2	5.5	5.0	5.1	4.4	2.8	3.7	3.5	2.8	3.6	4.1	5.0	4.2
SALT LAKE CITY	24	6.9	7.0	6.5	6.1	5.4	4.2	3.5	3.4	3.4	4.3	5.6	6.9	5.3
VT. BURLINGTON	49	7.2	6.9	6.6	6.7	6.7	6.1	5.8	5.7	6.0	6.5	7.9	7.8	6.7
VA. NORFOLK	40	5.6	5.2	5.1	4.8	4.8	4.9	5.0	5.0	4.6	4.4	4.6	5.4	5.0
RICHMOND	56	5.9	5.2	5.3	5.1	5.0	5.2	5.2	5.2	4.9	4.3	4.8	5.5	5.1
ROANOKE	11	6.7	6.3	6.4	6.3	6.3	5.8	6.0	5.9	5.5	5.2	5.3	6.2	6.0
WASH. NORTH HEAD	56	7.8	7.3	7.1	6.9	6.9	6.8	6.4	6.5	6.2	6.8	7.6	7.8	7.0
SEATTLE	24	8.0	7.7	7.4	6.9	6.4	6.4	4.9	5.3	5.6	7.2	8.0	8.1	6.8
SPOKANE	67	8.1	7.4	6.8	6.4	6.2	5.8	3.5	3.6	4.8	6.0	7.7	8.2	6.2
TATOOSH ISLAND	49	7.9	7.3	7.3	7.2	7.1	7.1	6.9	7.0	6.5	6.9	8.0	7.9	7.3
WALLA WALLA	50	8.1	7.4	6.3	5.4	5.0	4.5	2.4	2.7	4.0	5.2	7.4	8.4	5.6
YAKIMA	27	7.7	7.1	6.5	6.1	5.6	5.0	2.6	2.8	3.9	5.4	7.1	7.8	5.6
W.VA. ELKINS	27	7.9	7.6	7.4	7.0	6.8	6.7	6.5	6.4	6.1	6.1	6.9	7.6	6.9
PARKERSBURG	68	7.3	6.8	6.4	6.0	5.5	5.2	5.0	5.1	4.8	5.1	6.6	7.3	5.9
WIS. GREEN BAY	69	6.8	6.5	6.4	6.4	6.4	6.2	5.6	5.7	5.9	6.3	7.2	7.1	6.4
MILWAUKEE	58	6.5	6.3	6.3	6.0	5.7	5.5	4.5	4.8	5.1	5.5	6.5	6.7	5.8
WYO. CHEYENNE	56	4.8	5.6	5.7	6.2	6.3	5.3	4.8	5.0	4.5	4.4	5.0	5.1	5.2
LANDER	64	4.7	4.8	5.3	5.6	5.5	4.5	4.1	4.0	4.0	4.3	4.9	4.7	4.7
SHERIDAN	15	6.5	6.7	6.9	6.6	6.6	5.8	3.9	4.1	4.7	5.1	6.4	6.6	5.8
YELLOWSTONE	35	6.9	6.4	6.4	6.1	6.3	5.4	4.3	4.3	4.9	5.4	6.4	6.8	5.8
P.R. SAN JUAN	57	4.9	4.5	4.5	5.0	5.8	5.8	5.5	5.1	5.8	5.4	5.2	5.1	5.2
VIRGIN ISLANDS. ST. CROIX	6	5.6	5.8	5.7	5.9	7.2	6.9	7.0	6.1	7.0	6.5	6.0	5.8	6.3

* SUN BELOW HORIZON.

BASED ON PERIOD OF RECORD THROUGH DECEMBER 1959, EXCEPT IN A FEW INSTANCES. VALUES ARE IN TENTHS; (10.0 WOULD BE COMPLETE SKY COVERAGE). DERIVED FROM "NORMALS, MEANS, AND EXTREMES" TABLE IN U. S. WEATHER BUREAU PUBLICATION LOCAL CLIMATOLOGICAL DATA.

APPENDIX B
DESIGN OUTSIDE TEMPERATURES

OUTSIDE TEMPERATURES USED IN DESIGNING A HEATING SYSTEM*

State/City	Design Outside Temperature (°F)
ALABAMA	
Birmingham	10
Montgomery	10
ARIZONA	
Phoenix	25
Yuma	30
ARKANSAS	
Little Rock	5
CALIFORNIA	
Eureka	30
Fresno	25
Los Angeles	35
Sacramento	30
San Diego	35
San Francisco	35
COLORADO	
Denver	–10
Grand Junction	–15
CONNECTICUT	
Hartford	0
DISTRICT OF COLUMBIA	
Washington	0
GEORGIA	
Atlanta	10
IDAHO	
Boise	–10
Pocatello	– 5
ILLINOIS	
Cairo	0
Chicago	–10
Springfield	–10
INDIANA	
Fort Wayne	–10
Indianapolis	–10
IOWA	
Des Moines	–15
Dubuque	–20
Sioux City	–20

State/City	Design Outside Temperature (°F)
KANSAS	
Concordia	–10
Dodge City	–10
Wichita	–10
KENTUCKY	
Louisville	0
LOUISIANA	
New Orleans	20
Shreveport	20
MASSACHUSETTS	
Boston	0
MICHIGAN	
Alpena	–10
Detroit	–10
Grand Rapids	–10
Marquette	–10
Sault St. Marie	–20
MINNESOTA	
Duluth	–25
Minneapolis	–20
MISSISSIPPI	
Vicksburg	10
MISSOURI	
Kansas City	–10
St. Louis	0
MONTANA	
Havre	–30
Helena	–20
Kalispell	–20
NEBRASKA	
Lincoln	–10
North Platte	–20
NEVADA	
Reno	– 5
Winnemucca	–15
NEW HAMPSHIRE	
Concord	–15

*The design outside temperatures are approximately 15°F above the lowest temperature ever recorded by the meteorological station in the area.

State/City	Design Outside Temperature (°F)
NEW JERSEY	
Atlantic City	5
NEW MEXICO	
Albuquerque	0
Roswell	–10
NEW YORK	
Albany	–10
Binghamton	–10
Buffalo	– 5
New York City	0
Syracuse	–10
NORTH CAROLINA	
Asheville	0
Raleigh	10
NORTH DAKOTA	
Bismarck	–30
Devils Lake	–30
Fargo	–25
Williston	–35
OHIO	
Cincinnati	0
Cleveland	0
Columbus	–10
OKLAHOMA	
Oklahoma City	0
OREGON	
Portland	10
Roseburg	10
PENNSYLVANIA	
Harrisburg	0
Philadelphia	0
Pittsburgh	0
RHODE ISLAND	
Block Island	0
SOUTH CAROLINA	
Charleston	15
Columbia	10
SOUTH DAKOTA	
Huron	–20
Rapid City	–20
TENNESSEE	
Knoxville	0
Memphis	0
Nashville	0
TEXAS	
Abilene	15
Amarillo	–10
Austin	20
Brownsville	30
El Paso	10
Fort Worth	10
Galveston	20
San Antonio	20
UTAH	
Salt Lake City	–10
VERMONT	
Burlington	–10
VIRGINIA	
Norfolk	15
Richmond	15
WASHINGTON	
Seattle	15
Spokane	15
Tatoosh Island	15
Yakima	– 5
WEST VIRGINIA	
Elkins	–10
Parkersburg	–10
WISCONSIN	
Green Bay	–20
Madison	–15
Milwaukee	–15
WYOMING	
Cheyenne	–15
Lander	–18

APPENDIX C
CONVERSION FACTORS AND CONSTANTS

CONVERSION FACTORS

Length:	1 in. = 0.0833 ft = 2.54 cm
	1 micron = 3.281 x 10^{-6} ft = 10^{-4} cm
	1 angstrom unit = 10^{-8} cm
Heat/Power/Work:	1 Btu = 778.3 ft•lb = 252 cal
	1 kw•hr = 3413 Btu
	1 hp = 2544 Btu/hr
	1 ton of refrigeration = 12,000 Btu/hr
Heat Flux:	1 langley = 1 cal/cm^2
	1 cal/cm^2•min = 221.2 Btu/ft^2•hr
	1 watt/cm^2 = 3170 Btu/ft^2•hr
	1 langley/day = 3.687 Btu/ft^2•day
Pressure:	1 in. of water = 0.03613 lb/$in.^2$
	1 ft of water = 62.43 lb/ft^2
Thermal Conductivity:	1 cal/cm•sec•°C = 241.9 Btu/ft•hr•°F
	1 watt /cm•°C = 57.79 Btu/ft•hr•°F
Volume:	1 ft^3 = 7.481 gal (U.S.)

MISCELLANEOUS CONSTANTS

1 solar constant ≈2.0 langleys/min
≈442 Btu/ft^2•hr = 129.5 watts/ft^2
≈0.1394 watts/cm^2 = 1.394 kw/m^2

1 degree of latitude at 40° = 69 miles

Acceleration due to gravity at sea level = 32.17 ft/sec^2

Atmospheric pressure at sea level = 14.7 lb/$in.^2$

Stefan-Boltzmann constant = 0.173 x 10^{-8} Btu/ft^2•hr•°R^4

APPENDIX D
PHYSICAL PROPERTY DATA

THERMAL PROPERTY DATA

Substance	Temperature Range °F	Thermal Conductivity (k) Btu/hr•ft•°F	Average Density (ρ) lb/ft^3	Specific Heat (C_p) Btu/lb•°F
GASES				
Air	80	0.015	0.074*	0.24
	170	0.017	0.062*	0.24
	260	0.019	0.055*	0.24
	350	0.021	0.049*	0.24
LIQUIDS				
Water	32	0.32	62.4	1.0
	70	0.35	62.4	1.0
	140	0.38	62.4	1.0
	212	0.39	62.4	1.0
SOLIDS				
Building Materials:				
Brick	......	0.22 - 0.30	110 - 115	0.20 - 0.22
Concrete	......	0.50 - 0.75	120 - 145	0.15 - 0.25
Glass	......	0.42 - 0.50	150 - 175	0.20
Granite	......	1.08 - 2.33	165 - 172	0.20
Limestone	......	0.33 - 0.75	167 - 171	0.22
Sandstone	......	0.67 - 1.33	134 - 147	0.17 - 0.22
Insulating Materials:				
Corkboard	100	0.022 - 0.025	10.0	0.50
Fiberglass, batt	100	0.025 - 0.030	......	......
Glass wool, blanket	100	0.022 - 0.023	12.5	0.16
Insulating boards (Insulite, Celotex, Masonite, etc.)	100	0.027 - 0.031	14.8	......
Polystyrene (beadboard)	100	0.022	2.9	......
Polyurethane	100	0.011	5.0	......
Rock wool, loose	100	0.024 - 0.030	......	......
Metal Alloys:				
Aluminum	70 - 400	65 - 110	165	0.20 - 0.21
Copper	70 - 400	15 - 65	540	0.08 - 0.10
Nickel	70 - 400	8 - 12	537	0.10 - 0.11
Silver, cast	70 - 400	242	656	0.05 - 0.06
Steel (~1% C)	70 - 400	21 - 25	487	0.11
Zinc, cast	70 - 400	60 - 65	440	0.09 - 0.10
Other:				
Gravel, dry	......	......	90 - 120	......

*Based on sea-level pressure.

RADIATION PROPERTY DATA

Substance	Solar Absorptivity (α_S)	Emissivity (ϵ)
Aluminum:		
Highly polished	...	0.04 - 0.06
Rough, plate	...	0.06 - 0.07
Oxidized	...	0.2 - 0.6
Copper:		
Polished	...	0.02 - 0.03
Oxidized	...	0.5 - 0.8
Glass	0.14	0.88 - 0.94
Paint:		
Aluminum	...	0.20 - 0.40
Flat black	0.94	0.90 - 0.96

LINEAR EXPANSION DATA

Substance	Coefficient of Linear Expansion (δ) in./in./°F
Aluminum	$12 - 14 \times 10^{-6}$
Brick	5.3×10^{-6}
Concrete	8.0×10^{-6}
Copper	9.1×10^{-6}
Glass	5.0×10^{-6}
Granite	$4 - 5 \times 10^{-6}$
Limestone	$2 - 5 \times 10^{-6}$
Plastic:	
Plexiglas	41.0×10^{-6}
Polyethylene	$83 - 167 \times 10^{-6}$
Steel	6.3×10^{-6}

APPENDIX E
THEORETICAL COLLECTION SYSTEM CAPABILITY OF ISC BACKYARD SOLAR FURNACE

CONTENTS

SUN POSITION FOR VARIOUS NORTH LATITUDES

Summer Solstice (May-August)
Declination = +23.5 Degrees

Local Mean Sun Time AM	PM	Hour Angle (Deg)	30° N Alt	30° N Azi	35° N Alt	35° N Azi	40° N Alt	40° N Azi	45° N Alt	45° N Azi
6	6	90	11.5	110.6	13.2	109.6	14.9	108.4	16.4	107.1
7	5	75	23.9	104.3	25.0	102.2	26.0	99.7	26.7	97.5
8	4	60	36.6	98.4	37.2	94.4	37.4	91.3	37.3	86.8
9	3	45	49.5	86.8	49.5	86.8	48.9	80.6	47.8	74.9
10	2	30	62.5	83.2	61.6	74.6	59.8	65.7	57.5	58.6
11	1	15	75.1	67.4	72.6	52.5	69.2	41.9	65.3	34.6
Noon		0	83.5	0.0	78.5	0.0	73.5	0.0	68.5	0.0

Equinoxes: Spring (February-May) & Fall (August-November)
Declination = 0 Degrees

Local Mean Sun Time AM	PM	Hour Angle (Deg)	30° N Alt	30° N Azi	35° N Alt	35° N Azi	40° N Alt	40° N Azi	45° N Alt	45° N Azi
6	6	90	0.0	..	0.0	..	0.0	..	0.0	..
7	5	75	13.0	82.5	12.2	81.2	11.4	80.2	10.5	79.2
8	4	60	25.7	74.0	24.2	71.7	22.5	69.6	20.7	67.8
9	3	45	37.8	63.5	35.4	60.2	32.8	57.3	30.0	54.7
10	2	30	48.6	49.1	45.2	45.2	41.6	42.0	37.8	39.3
11	1	15	56.8	28.2	52.3	25.0	47.7	22.6	43.1	20.8
Noon		0	60.0	0.0	55.0	0.0	50.0	0.0	45.0	0.0

Winter Solstice (November-February)
Declination = −23.5 Degrees

Local Mean Sun Time AM	PM	Hour Angle (Deg)	30° N Alt	30° N Azi	35° N Alt	35° N Azi	40° N Alt	40° N Azi	45° N Alt	45° N Azi
6	6	90	0.0	..	0.0	..	0.0	..	0.0	..
7	5	75	0.4	62.4	0.0	62.4	0.0	62.4	0.0	62.4
8	4	60	11.4	54.1	8.4	53.4	5.4	52.9	2.4	52.6
9	3	45	21.2	44.1	17.6	42.9	13.9	41.9	10.2	41.2
10	2	30	29.2	31.7	25.0	30.4	20.6	29.3	16.2	28.5
11	1	15	34.6	16.8	29.8	15.9	25.0	15.2	20.1	14.6
Noon		0	36.5	0.0	31.5	0.0	26.5	0.0	21.5	0.0

Note: $\sin(\text{altitude}) = \cos D \cos H \cos L + \sin D \sin L$

$\sin(\text{azimuth}) = \dfrac{\cos D \sin H}{\cos(\text{altitude})}$

where D = declination, deg
H = hour angle, deg
L = latitude, deg

BOUNDARY CONDITIONS

Collector Unit:

ISC solar furnace (collector surface inclined 60° to the horizontal) $\psi = 60°$

Orientation:

Latitude — 30°, 35°, 40° and 45° North

Longitude — standard meridians where local and sun times coincide

Collector faces due south — $\beta = 0°$

Altitude — sea level (Note: Altitude does not significantly affect the total daily shortwave radiation.)

Weather Conditions:

Clear

EQUATIONS AND DEFINITIONS FOR SHORTWAVE RADIATION CALCULATIONS

Equations:

$$G_n = \frac{A_o C}{e^{(B/\sin X)}}$$

$$\cos i = .5 \cos Z + .866 \sin Z \cos A$$

$$\cos j = \cos \phi \cos Z - \sin \phi \sin Z \cos A$$

$$\cos k = \cos (120° - 2\phi) \cos Z + \sin (120° - 2\phi) \sin Z \cos A$$

$$G_i = G_n \cos i$$

$$G_l = \tau_i G_i$$

$$G_h = G_n \cos Z$$

$$G_d = .75\, G_h/\epsilon \qquad \text{(based on viewing factor to open sky of 3/4)}$$

$$G_D = .605\, G_d$$

$$G_m = G_n \cos j$$

$$F_a = 1 - \tfrac{1}{3}\left\{(1 + [\cos (60° + \phi)]/f_x)\frac{\tan A}{\cos \phi}\right\} \qquad \text{for } F_a > 0.5$$

$F_a = .75 f_x \cos\phi/\tan A\ [\cos(60° + \phi) + f_x]$ for $F_a < 0.5$

where:

$f_x = \sin(60° + 2\phi - X_s)/\sin(X_s - \phi)$

$F_x = \sin(60° + 2\phi - X_s)/\sin(X_s - \phi)$ for $30° + 1.5\phi < X_s < 60° + 2\phi$

$F_x = 1$ for $X_s < 30° + 1.5\phi$

where:

$X_s = \tan^{-1}(\tan X/\cos A)$

$G_r = \rho\ F_a F_x G_m$

$G_R = \tau_k G_r$

Definitions:

A = solar azimuth angle, deg

A_o = apparent extraterrestrial irradiation at air mass 0, $Btu/ft^2{\cdot}hr$

B = atmospheric extinction coefficient, dimensionless

C = clearness number, dimensionless

F_a, F_x = mirror configuration factors for solar azimuth and altitude, respectively, dimensionless (Note: F_aF_x is the fraction of the direct solar radiation that is reflected onto the collector surface from the mirror surface accounting for the radiation reflected to the side and over the top of the collector.)

G_d = diffuse sky radiation, $Btu/ft^2{\cdot}hr$ (Note: G_d is independent of the collector surface inclination, except for the view that the collector surface has of the open sky.)

G_D = diffuse sky radiation transmitted through the collector covers, $Btu/ft^2{\cdot}hr$

G_h = direct solar radiation incident on a horizontal surface, $Btu/ft^2{\cdot}hr$

G_i = direct solar radiation incident on the collector surface, $Btu/ft^2{\cdot}hr$

G_I = direct solar radiation transmitted through the collector covers, $Btu/ft^2{\cdot}hr$

G_m = direct solar radiation incident on the mirror surface, $Btu/ft^2{\cdot}hr$

G_n = direct solar radiation incident on a surface at sea level that is placed normal to the sun's rays, $Btu/ft^2{\cdot}hr$

G_o = solar constant (Note: G_o varies with solar activity from 401 to 456 $Btu/ft^2{\cdot}hr$. A typical value of 442 is used in this book.)

G_r = direct solar radiation reflected onto the collector surface from the mirror surface, Btu/ft²•hr

G_R = direct solar radiation from mirror reflections that is transmitted through the collector covers, Btu/ft²•hr

i = angle of incidence of sunlight on collector, deg

j = angle of incidence of sunlight on mirror, deg

k = angle of incidence of sunlight reflected onto collector, deg

X = solar altitude angle, deg

Z = solar zenith angle, deg (Note: Z is the angle that the sun's rays make with the vertical, i.e., 90° − X)

β = orientation angle of the collector in the horizontal plane with respect to south, deg (Note: β is 0° for these calculations.)

ϵ = ratio of direct solar radiation to diffuse sky radiation on a horizontal surface, dimensionless (Note: See Figure 2 for ϵ values.)

ρ = reflectance of mirror. A value of 0.95 is used in this book.

τ_i, τ_k = transmittance factors for direct solar radiation through the collector covers, dimensionless (Note: τ_i is the transmitted fraction of that incident on the collector surface; τ_k is the transmitted fraction of that reflected onto the collector surface from the mirror surface. See Figure 1 for τ values.)

ϕ = inclination angle of the mirror surface, deg

ψ = tilt of a surface, measured from the horizontal, deg

Notes: 1) Total flux per hour = $G_I + G_D + G_R$

2) Total flux per day = $\Sigma(G_I + G_D + G_R)$

3) Percentage improvement = $G_R / (G_I + G_D)$

4) The general equation for angle of incidence of sunlight onto a flat surface is $i = \cos^{-1} [\cos Z \cos\psi + \sin Z \sin\psi \cos (A - \beta)]$

Total Daily Shortwave Radiation (Direct & Diffuse) Transmitted Through the ISC Solar Furnace Covers at 30° North Latitude on a Cloudless Summer Solstice Day[1]

Local Mean Sun Time		Sun Position (Deg)			Collector only										
					Direct Solar Radiation						Diffuse Sky Radiation				Total
AM	PM	Azi	Alt	Zen	G_n	cos i	G_i	i	τ_i [2]	G_I	G_h	ϵ [3]	G_d	G_D [4]	$G_I + G_D$
6	6	110.6	11.5	78.5	153.6	0.0	0.0	—	0.0	0.0	30.6	1.55	13.2	8.0	8.0
7	5	104.3	23.9	66.1	262.7	0.007	1.8	—	0.0	0.0	106.4	2.60	27.3	16.5	16.5
8	4	98.4	36.6	53.4	310.4	0.197	61.0	—	0.0	0.0	185.1	3.60	34.3	20.8	20.8
9	3	86.8	49.5	40.5	335.0	0.412	137.9	87.0	0.08	11.0	254.8	4.52	37.6	22.7	33.7
10	2	83.2	62.5	27.5	348.5	0.491	171.1	84.3	0.14	24.0	309.2	5.30	38.9	23.5	47.5
11	1	67.4	75.1	14.9	355.4	0.569	202.1	74.3	0.43	86.9	343.5	5.77	39.7	24.0	110.9
Noon		0.0	83.5	6.5	357.5	0.595	212.6	53.5	0.71	150.9	355.2	5.98	39.6	24.0	174.9
					Total (Btu/ft²•day) = 394.7									255.0	649.7

	Mirror Contribution (Direct Solar Radiation only)[5]																	
	0° Inclination						10° Inclination						20° Inclination					
F_a	G_m	F_x	G_r [6]	k	τ_k	G_R	G_m	F_x	G_r [6]	k	τ_k	G_R	G_m	F_x	G_r [6]	k	τ_k	G_R
0.0	30.6	1.00	0.0	—	0.0	0.0	4.0	1.00	0.0	—	0.0	0.0	0.0	1.00	0.0	—	0.0	0.0
0.0	106.4	1.00	0.0	—	0.0	0.0	63.1	1.00	0.0	—	0.0	0.0	17.9	1.00	0.0	—	0.0	0.0
0.0	185.1	0.67	0.0	—	0.0	0.0	139.0	1.00	0.0	—	0.0	0.0	88.7	1.00	0.0	—	0.0	0.0
0.08	254.8	0.24	4.6	89.4	0.02	0.1	213.1	0.80	13.0	88.4	0.05	0.7	165.0	1.00	12.5	87.5	0.07	0.9
0.18	309.2	0.0	0.0	—	0.0	0.0	276.5	0.38	18.0	88.0	0.05	0.9	235.4	0.90	36.2	85.9	0.11	4.0
0.60	343.5	0.0	0.0	—	0.0	0.0	322.4	0.09	16.5	88.1	0.05	0.8	291.5	0.51	84.7	80.7	0.24	20.3
1.00	355.2	0.0	0.0	—	0.0	0.0	342.8	0.0	0.0	—	0.0	0.0	319.9	0.32	97.2	73.5	0.45	43.7
Total (Btu/ft²•day) = 0.2												4.8						94.1
Percentage Improvement (%) = 0.0												0.7						14.5

Total Shortwave Radiation[7] $G_I + G_D + G_R$		
0° Inclination	10° Inclination	20° Inclination
8.0	8.0	8.0
16.5	16.5	16.5
20.8	20.8	20.8
33.8	34.4	34.6
47.5	48.4	51.5
110.9	111.7	131.2
174.9	174.9	218.6
649.9[8]	654.5[8]	743.8[8]

Notes: 1) Declination = +23.5 degrees.
2) Interpolated from Figure 1.
3) Interpolated from Figure 2.
4) Based on an effective transmissivity of 0.605 through two cover glasses.
5) The reflected diffuse radiation was neglected, because it is estimated to be less than 5 percent of the diffuse radiation incident on the collector surface, i.e., $< 0.05\ G_d$.
6) Based on a mirror surface reflectivity of 0.95.
7) The reduction in shortwave radiation to the collector that results from mirror shadowing was neglected.
8) Total flux per day (Btu/ft²·day).

Total Daily Shortwave Radiation (Direct & Diffuse) Transmitted Through the ISC Solar Furnace Covers at 30° North Latitude on a Cloudless Spring or Fall Equinox Day[1]

Local Mean Sun Time		Sun Position (Deg)			Collector only										
					Direct Solar Radiation						Diffuse Sky Radiation				Total
AM	PM	Azi	Alt	Zen	G_n	cos i	G_i	i	τ_i[2]	G_I	G_h	ϵ[3]	G_d	G_D[4]	$G_I + G_D$
6	6	—	0.0	90.0	0.0	0.0	0.0	—	0.0	0.0	0.0	0.0	0.0	0.0	0.0
7	5	82.5	13.0	77.0	173.2	0.223	38.6	82.8	0.18	6.9	39.0	1.67	15.6	9.4	16.3
8	4	74.0	25.7	64.3	271.9	0.432	117.4	74.0	0.44	51.7	117.9	2.75	28.6	17.3	69.0
9	3	63.5	37.8	52.2	313.4	0.612	191.7	63.8	0.63	120.8	192.1	3.67	34.9	21.1	141.9
10	2	49.1	48.6	41.4	333.7	0.753	251.4	51.6	0.73	183.5	250.3	4.45	37.5	22.7	206.2
11	1	28.2	56.8	33.2	343.6	0.836	287.4	38.1	0.76	218.4	287.5	5.00	38.4	23.2	241.6
Noon		0.0	60.0	30.0	346.5	0.866	300.1	30.0	0.76	228.1	300.1	5.17	38.7	23.4	251.5
					Total (Btu/ft²·day) =					1390.7				210.8	1601.5

Mirror Contribution (Direct Solar Radiation only)[5]																		
	0° Inclination						10° Inclination						20° Inclination					
F_a	G_m	F_x	G_r[6]	k	τ_k	G_R	G_m	F_x	G_r[6]	k	τ_k	G_R	G_m	F_x	G_r[6]	k	τ_k	G_R
0.0	0.0	1.00	0.0	—	0.0	0.0	0.0	1.00	0.0	—	0.0	0.0	0.0	1.00	0.0	—	0.0	0.0
0.20	39.0	1.00	7.4	84.5	0.14	1.0	9.1	1.00	1.7	83.1	0.19	0.3	0.0	1.00	0.0	82.5	0.19	0.0
0.43	117.9	1.00	48.2	81.1	0.22	10.6	73.6	1.00	30.1	77.1	0.33	9.9	27.0	1.00	11.0	74.6	0.42	4.6
0.67	192.1	0.62	75.8	80.3	0.25	19.0	146.2	1.00	93.1	72.6	0.48	44.7	95.8	1.00	61.0	66.8	0.59	36.0
0.81	250.3	0.26	50.1	82.6	0.19	9.5	208.2	0.84	134.6	70.1	0.53	71.3	159.7	1.00	122.9	58.8	0.68	83.6
0.91	287.5	0.07	17.4	87.2	0.08	1.4	250.5	0.54	116.9	69.7	0.53	62.0	205.8	1.00	177.9	52.9	0.72	128.1
1.00	300.1	0.0	0.0	—	0.0	0.0	265.4	0.45	113.5	70.0	0.53	60.2	222.7	1.00	211.6	50.0	0.73	154.5
Total (Btu/ft²·day) =						83.0						436.6						659.1
Percentage Improvement (%) =						5.1						27.3						41.2

Total Shortwave Radiation[7] $G_I + G_D + G_R$		
0° Inclination	10° Inclination	20° Inclination
0.0	0.0	0.0
17.3	16.6	16.3
79.6	78.9	73.6
160.9	186.6	177.9
215.7	277.5	289.8
243.0	303.6	359.7
251.5	311.7	406.0
1684.5[8]	2038.1[8]	2260.6[8]

Notes:
1) Declination = 0 degrees.
2) Interpolated from Figure 1.
3) Interpolated from Figure 2.
4) Based on an effective transmissivity of 0.605 through two cover glasses.
5) The reflected diffuse radiation was neglected, because it is estimated to be less than 5 percent of the diffuse radiation incident on the collector surface, i.e., $< 0.05\ G_d$.
6) Based on a mirror surface reflectivity of 0.95.
7) The reduction in shortwave radiation to the collector that results from mirror shadowing was neglected.
8) Total flux per day (Btu/ft²·day).

Total Daily Shortwave Radiation (Direct & Diffuse) Transmitted Through the ISC Solar Furnace Covers at 30° North Latitude on a Cloudless Winter Solstice Day[1]

Local Mean Sun Time		Sun Position (Deg)			Collector only										Total
					Direct Solar Radiation						Diffuse Sky Radiation				
AM	PM	Azi	Alt	Zen	G_n	cos i	G_i	i	τ_i[2]	G_I	G_h	ϵ[3]	G_d	G_D[4]	$G_I + G_D$
6	6	—	0.0	90.0	0.0	0.0	0.0	—	0.0	0.0	0.0	0.0	0.0	0.0	0.0
7	5	62.4	0.4	89.6	0.0	0.492	0.0	66.2	0.60	0.0	0.0	0.05	0.0	0.0	0.0
8	4	54.1	11.4	78.6	152.2	0.607	92.3	56.2	0.71	65.5	30.1	1.50	13.4	8.1	73.6
9	3	44.1	21.2	68.8	246.8	0.803	198.1	44.8	0.74	146.6	89.2	2.40	24.8	15.0	161.6
10	2	31.7	29.2	60.8	287.0	0.887	254.6	31.7	0.76	193.5	140.0	3.02	30.9	18.7	212.2
11	1	16.8	34.6	55.4	305.0	0.966	294.7	17.4	0.77	226.9	173.2	3.42	33.8	20.4	247.3
Noon		0.0	36.5	53.5	310.1	0.994	308.1	6.5	0.77	237.2	184.5	3.57	34.5	20.9	258.1
									Total (Btu/ft²•day) =	1502.2				145.3	1647.5

	Mirror Contribution (Direct Solar Radiation only)[5]																	
	0° Inclination						10° Inclination						20° Inclination					
F_a	G_m	F_x	G_r[6]	k	τ_k	G_R	G_m	F_x	G_r[6]	k	τ_k	G_R	G_m	F_x	G_r[6]	k	τ_k	G_R
0.0	0.0	1.00	0.0	—	0.0	0.0	0.0	1.00	0.0	—	0.0	0.0	0.0	1.00	0.0	—	0.0	0.0
0.68	0.0	1.00	0.0	66.4	0.59	0.0	0.0	1.00	0.0	62.9	0.64	0.0	0.0	1.00	0.0	62.8	0.64	0.0
0.77	30.1	1.00	22.0	63.9	0.63	13.9	3.7	1.00	2.7	56.9	0.69	1.9	0.0	1.00	0.0	54.1	0.71	0.0
0.84	89.2	1.00	71.2	63.3	0.63	44.8	47.9	1.00	38.2	52.1	0.72	27.5	5.2	1.00	4.1	45.2	0.74	3.0
0.90	140.0	1.00	119.7	64.2	0.62	74.2	94.4	1.00	80.7	48.8	0.73	58.9	45.9	1.00	39.2	36.5	0.76	29.8
0.95	173.2	0.76	118.8	65.8	0.60	71.3	127.0	1.00	114.6	47.0	0.74	84.8	76.9	1.00	69.4	29.5	0.76	52.7
1.00	184.5	0.67	117.4	66.5	0.59	69.3	138.4	1.00	131.5	46.5	0.74	97.3	88.1	1.00	83.7	26.5	0.77	64.4
Total (Btu/ft²•day) =						477.7						443.5						235.4
Percentage Improvement (%) =						29.0						26.9						14.3

Total Shortwave Radiation[7] $G_I + G_D + G_R$		
0° Inclination	10° Inclination	20° Inclination
0.0	0.0	0.0
0.0	0.0	0.0
87.5	75.5	73.6
206.4	189.1	164.6
286.4	271.1	242.0
318.6	332.1	300.0
327.4	355.4	322.5
2125.2[8]	2091.0[8]	1882.9[8]

Notes:
1) Declination = − 23.5 degrees.
2) Interpolated from Figure 1.
3) Interpolated from Figure 2.
4) Based on an effective transmissivity of 0.605 through two cover glasses.
5) The reflected diffuse radiation was neglected, because it is estimated to be less than 5 percent of the diffuse radiation incident on the collector surface, i.e., $< 0.05\ G_d$.
6) Based on a mirror surface reflectivity of 0.95.
7) The reduction in shortwave radiation to the collector that results from mirror shadowing was neglected.
8) Total flux per day (Btu/ft²·day).

Total Daily Shortwave Radiation (Direct & Diffuse) Transmitted Through the ISC Solar Furnace Covers at 35° North Latitude on a Cloudless Summer Solstice Day[1]

Local Mean Sun Time		Sun Position (Deg)			Collector only										
					Direct Solar Radiation						Diffuse Sky Radiation				Total
AM	PM	Azi	Alt	Zen	G_n	cos i	G_i	i	τ_i[2]	G_I	G_h	ϵ[3]	G_d	G_D[4]	$G_I + G_D$
6	6	109.6	13.2	76.8	175.7	0.0	0.0	—	0.0	0.0	40.1	1.67	16.0	9.7	9.7
7	5	102.2	25.0	65.0	268.5	0.045	12.2	—	0.0	0.0	113.5	2.69	28.1	17.0	17.0
8	4	94.4	37.2	52.8	311.9	0.249	77.8	—	0.0	0.0	188.6	3.62	34.7	21.0	21.0
9	3	86.8	49.5	40.5	335.0	0.412	137.9	87.0	0.08	11.0	254.8	4.52	37.6	22.7	33.7
10	2	74.6	61.6	28.4	347.3	0.549	191.0	76.9	0.33	63.0	306.0	5.26	38.8	23.5	86.5
11	1	52.5	72.6	17.4	354.4	0.635	225.0	63.4	0.63	141.8	338.2	5.70	39.6	24.0	165.8
Noon		0.0	78.5	11.5	356.5	0.663	236.2	48.5	0.74	174.8	349.3	5.86	39.8	24.1	198.9
					Total (Btu/ft²•day) = 606.4									259.9	866.3

	Mirror Contribution (Direct Solar Radiation only)[5]																	
	0° Inclination[4]						10° Inclination						20° Inclination					
F_a	G_m	F_x	G_r[6]	k	τ_k	G_R	G_m	F_x	G_r[6]	k	τ_k	G_R	G_m	F_x	G_r[6]	k	τ_k	G_R
0.0	40.1	1.00	0.0	—	0.0	0.0	9.8	1.00	0.0	—	0.0	0.0	0.0	1.00	0.0	—	0.0	0.0
0.0	113.5	1.00	0.0	—	0.0	0.0	69.5	1.00	0.0	—	0.0	0.0	23.4	1.00	0.0	—	0.0	0.0
0.0	188.6	0.64	0.0	—	0.0	0.0	142.6	1.00	0.0	—	0.0	0.0	92.2	1.00	0.0	—	0.0	0.0
0.08	254.8	0.24	4.6	89.4	0.02	0.1	213.1	0.80	13.0	88.4	0.04	0.5	165.0	1.00	12.5	87.5	0.07	0.9
0.41	306.0	0.0	0.0	—	0.0	0.0	272.6	0.40	42.5	85.2	0.12	5.1	230.9	0.94	84.5	80.5	0.24	20.3
0.78	338.2	0.0	0.0	—	0.0	0.0	314.6	0.15	35.0	85.5	0.11	3.9	281.5	0.58	121.0	73.7	0.44	53.2
1.00	349.3	0.0	0.0	—	0.0	0.0	331.7	0.03	9.5	88.5	0.04	0.4	304.0	0.43	124.2	68.5	0.56	69.6
Total (Btu/ft²•day) =						0.2						19.4						218.4
Percentage Improvement (%) =						0.0						2.2						25.2

Total Shortwave Radiation[7] $G_I + G_D + G_R$		
0° Inclination	10° Inclination	20° Inclination
9.7	9.7	9.7
17.0	17.0	17.0
21.0	21.0	21.0
33.8	34.2	34.6
86.5	91.6	106.8
165.8	169.7	219.0
198.9	199.3	268.5
866.5[8]	885.7[8]	1084.7[8]

Notes:
1) Declination = + 23.5 degrees.
2) Interpolated from Figure 1.
3) Interpolated from Figure 2.
4) Based on an effective transmissivity of 0.605 through two cover glasses.
5) The reflected diffuse radiation was neglected, because it is estimated to be less than 5 percent of the diffuse radiation incident on the collector surface, i.e., $< 0.05\ G_d$.
6) Based on a mirror surface reflectivity of 0.95.
7) The reduction in shortwave radiation to the collector that results from mirror shadowing was neglected.
8) Total flux per day (Btu/ft²·day).

Total Daily Shortwave Radiation (Direct & Diffuse) Transmitted Through the ISC Solar Furnace Covers at 35° North Latitude on a Cloudless Spring or Fall Equinox Day[1]

Local Mean Sun Time		Sun Position (Deg)			Collector only										
					Direct Solar Radiation						Diffuse Sky Radiation				Total
AM	PM	Azi	Alt	Zen	G_n	cos i	G_i	i	τ_i[2]	G_I	G_h	ϵ[3]	G_d	G_D[4]	$G_I + G_D$
6	6	—	0.0	90.0	0.0	0.0	0.0	—	0.0	0.0	0.0	0.0	0.0	0.0	0.0
7	5	81.2	12.2	77.8	163.1	0.235	38.4	81.6	0.21	8.1	34.5	1.60	14.4	8.7	16.8
8	4	71.7	24.2	65.8	264.3	0.453	119.7	71.8	0.49	58.7	108.4	2.63	27.5	16.6	75.3
9	3	60.2	35.4	54.6	307.2	0.640	196.8	60.3	0.67	131.9	178.0	3.48	34.1	20.6	152.5
10	2	45.2	45.2	44.8	328.4	0.785	257.7	47.2	0.74	190.7	233.0	4.22	36.8	22.3	213.0
11	1	25.0	52.3	37.7	338.7	0.876	296.6	33.0	0.76	225.4	268.0	4.71	37.9	22.9	248.3
Noon		0.0	55.0	35.0	341.7	0.906	309.7	25.0	0.77	238.5	279.9	4.90	38.1	23.1	261.6
										Total (Btu/ft²·day) = 1468.1				205.3	1673.4

Mirror Contribution (Direct Solar Radiation only)[5]																		
	0° Inclination						10° Inclination						20° Inclination					
F_a	G_m	F_x	G_r[6]	k	τ_k	G_R	G_m	F_x	G_r[6]	k	τ_k	G_R	G_m	F_x	G_r[6]	k	τ_k	G_R
0.0	0.0	1.00	0.0	—	0.0	0.0	0.0	1.00	0.0	—	0.0	0.0	0.0	1.00	0.0	—	0.0	0.0
0.23	34.5	1.00	7.5	83.5	0.17	1.3	6.3	1.00	1.4	81.9	0.21	0.3	0.0	1.00	0.0	81.2	0.23	0.0
0.50	108.4	1.00	51.5	79.4	0.29	14.9	64.8	1.00	30.8	74.9	0.41	12.6	19.4	1.00	9.2	72.3	0.48	4.4
0.71	178.0	0.72	86.4	78.1	0.31	26.8	131.8	1.00	88.9	69.6	0.54	48.0	81.6	1.00	55.0	63.3	0.63	34.7
0.83	233.0	0.36	66.1	79.6	0.28	18.5	189.3	0.99	147.8	66.3	0.59	87.2	139.8	1.00	110.2	54.8	0.71	78.2
0.92	268.0	0.17	39.8	83.0	0.18	7.2	227.9	0.69	137.4	65.1	0.61	83.8	181.0	1.00	158.2	47.9	0.73	115.5
1.00	279.9	0.11	29.2	85.0	0.13	3.8	241.6	0.60	137.7	65.0	0.61	84.0	196.0	1.00	186.2	45.0	0.74	137.8
Total (Btu/ft²·day) =						141.2						547.8						603.4
Percentage Improvement (%) =						8.4						32.7						36.1

Total Shortwave Radiation[7] $G_I + G_D + G_R$		
0° Inclination	10° Inclination	20° Inclination
0.0	0.0	0.0
18.1	17.1	16.8
90.2	87.9	79.7
179.3	200.5	187.2
231.5	300.2	291.2
255.5	332.1	363.8
265.4	345.6	399.4
1814.6[8]	2221.2[8]	2276.8[8]

Notes:
1) Declination = 0 degrees.
2) Interpolated from Figure 1.
3) Interpolated from Figure 2.
4) Based on an effective transmissivity of 0.605 through two cover glasses.
5) The reflected diffuse radiation was neglected, because it is estimated to be less than 5 percent of the diffuse radiation incident on the collector surface, i.e., $< 0.05\ G_d$.
6) Based on a mirror surface reflectivity of 0.95.
7) The reduction in shortwave radiation to the collector that results from mirror shadowing was neglected.
8) Total flux per day (Btu/ft²·day).

Total Daily Shortwave Radiation (Direct & Diffuse) Transmitted Through the ISC Solar Furnace Covers at 35° North Latitude on a Cloudless Winter Solstice Day[1]

Local Mean Sun Time		Sun Position (Deg)			Collector only										
					Direct Solar Radiation						Diffuse Sky Radiation				Total
AM	PM	Azi	Alt	Zen	G_n	cos i	G_i	i	τ_i[2]	G_I	G_h	ϵ[3]	G_d	G_D[4]	$G_I + G_D$
6	6	—	0.0	90.0	0.0	0.0	0.0	—	0.0	0.0	0.0	0.0	0.0	0.0	0.0
7	5	62.4	0.0	90.0	0.0	0.0	0.0	66.3	0.60	0.0	0.0	0.0	0.0	0.0	0.0
8	4	53.4	8.4	81.6	104.5	0.584	61.0	56.3	0.71	43.3	15.3	1.23	8.3	5.0	48.3
9	3	42.9	17.6	72.4	220.2	0.756	166.4	44.3	0.74	123.1	66.6	2.10	21.1	12.8	135.9
10	2	30.4	25.0	65.0	268.5	0.888	238.5	30.8	0.76	181.3	113.5	2.70	28.0	16.9	198.2
11	1	15.9	29.8	60.2	289.3	0.971	281.0	15.9	0.77	216.4	143.8	3.07	31.2	18.9	235.3
Noon		0.0	31.5	58.5	295.3	0.9997	295.2	1.5	0.77	227.3	154.3	3.20	32.2	19.5	246.8
					Total (Btu/ft²•day) = 1355.5									126.7	1482.2

Mirror Contribution (Direct Solar Radiation only)[5]																		
	0° Inclination						10° Inclination						20° Inclination					
F_a	G_m	F_x	G_r[6]	k	τ_k	G_R	G_m	F_x	G_r[6]	k	τ_k	G_R	G_m	F_x	G_r[6]	k	τ_k	G_R
0.0	0.0	1.00	0.0	—	0.0	0.0	0.0	1.00	0.0	—	0.0	0.0	0.0	1.00	0.0	—	0.0	0.0
0.68	0.0	1.00	0.0	—	0.0	0.0	0.0	1.00	0.0	—	0.0	0.0	0.0	1.00	0.0	—	0.0	0.0
0.78	15.3	1.00	11.3	62.1	0.65	7.1	0.0	1.00	0.0	55.5	0.70	0.0	0.0	1.00	0.0	53.4	0.71	0.0
0.85	66.6	1.00	53.8	60.4	0.66	35.5	29.1	1.00	23.5	49.5	0.73	17.2	0.0	1.00	0.0	43.4	0.74	0.0
0.90	113.5	1.00	97.0	60.3	0.66	64.0	69.5	1.00	59.4	45.0	0.74	44.0	23.4	1.00	20.0	33.6	0.76	15.2
0.95	143.8	1.00	129.8	61.1	0.66	85.7	98.0	1.00	88.4	42.4	0.75	66.3	49.2	1.00	44.4	25.2	0.77	34.2
1.00	154.3	0.91	133.4	61.5	0.65	86.7	108.2	1.00	102.8	41.5	0.75	77.1	58.9	1.00	56.0	21.5	0.77	43.1
Total (Btu/ft²•day) =						471.9						332.1						141.9
Percentage Improvement (%) =						31.8						22.4						9.6

Total Shortwave Radiation[7] $G_I + G_D + G_R$		
0° Inclination	10° Inclination	20° Inclination
0.0	0.0	0.0
0.0	0.0	0.0
55.7	48.3	48.3
171.4	153.1	135.9
262.2	242.2	213.4
321.0	301.6	269.5
333.5	323.9	289.9
1954.1[8]	1814.3[8]	1624.1[8]

Notes: 1) Declination = − 23.5 degrees.
2) Interpolated from Figure 1.
3) Interpolated from Figure 2.
4) Based on an effective transmissivity of 0.605 through two cover glasses.
5) The reflected diffuse radiation was neglected, because it is estimated to be less than 5 percent of the diffuse radiation incident on the collector surface, i.e., $< 0.05\ G_d$.
6) Based on a mirror surface reflectivity of 0.95.
7) The reduction in shortwave radiation to the collector that results from mirror shadowing was neglected.
8) Total flux per day (Btu/ft²·day).

Total Daily Shortwave Radiation (Direct & Diffuse) Transmitted Through the ISC Solar Furnace Covers at 40° North Latitude on a Cloudless Summer Solstice Day[1]

Local Mean Sun Time		Sun Position (Deg)			Collector only										Total
					Direct Solar Radiation						Diffuse Sky Radiation				
AM	PM	Azl	Alt	Zen	G_n	cos i	G_i	i	τ_i[2]	G_I	G_h	ϵ[3]	G_d	G_D[4]	$G_I + G_D$
6	6	108.4	14.9	75.1	194.8	0.0	0.0	—	0.0	0.0	50.1	1.85	18.1	11.0	11.0
7	5	99.7	26.0	64.0	273.3	0.088	24.1	—	0.0	0.0	119.8	2.76	29.0	17.5	17.5
8	4	91.3	37.4	52.6	312.4	0.288	90.0	—	0.0	0.0	189.8	3.65	34.7	21.0	21.0
9	3	80.6	48.9	41.1	334.2	0.470	157.0	81.1	0.22	34.5	251.8	4.46	37.7	22.8	57.3
10	2	65.7	59.8	30.2	346.4	0.611	211.8	69.1	0.56	118.6	299.4	5.17	38.6	23.4	142.0
11	1	41.9	69.2	20.8	352.8	0.696	245.7	54.8	0.71	174.4	329.8	5.60	39.3	23.8	198.2
Noon		0.0	73.5	16.5	354.8	0.725	257.4	43.5	0.75	193.0	340.2	5.73	39.6	24.0	217.0
					Total (Btu/ft²•day) = 848.0									263.0	1111.0

Mirror Contribution (Direct Solar Radiation only)[5]

F_a	0° Inclination						10° Inclination						20° Inclination					
	G_m	F_x	G_r[6]	k	τ_k	G_R	G_m	F_x	G_r[6]	k	τ_k	G_R	G_m	F_x	G_r[6]	k	τ_k	G_R
0.0	50.1	1.00	0.0	—	0.0	0.0	16.6	1.00	0.0	—	0.0	0.0	0.0	1.00	0.0	—	0.0	0.0
0.0	119.8	1.00	0.0	—	0.0	0.0	75.3	1.00	0.0	—	0.0	0.0	28.6	1.00	0.0	—	0.0	0.0
0.0	189.8	0.63	0.0	—	0.0	0.0	143.8	1.00	0.0	—	0.0	0.0	93.4	1.00	0.0	—	0.0	0.0
0.25	251.8	0.26	15.5	88.2	0.04	0.6	209.9	0.82	40.9	85.2	0.12	4.9	161.5	1.00	38.4	82.7	0.18	6.9
0.63	299.4	0.0	0.0	89.9	0.0	0.0	264.6	0.45	71.3	81.8	0.21	15.0	221.7	1.00	132.7	74.6	0.42	55.7
0.85	329.8	0.0	0.0	—	0.0	0.0	303.0	0.22	53.8	82.0	0.21	11.3	267.1	0.68	146.7	67.6	0.57	83.6
1.00	340.2	0.0	0.0	—	0.0	0.0	317.5	0.13	39.2	83.5	0.17	6.7	285.2	0.56	151.7	63.5	0.63	95.6
Total (Btu/ft²•day) =						1.2						69.1						388.0
Percentage Improvement (%) =						0.1						6.2						34.9

Total Shortwave Radiation[7] $G_I + G_D + G_R$		
0° Inclination	10° Inclination	20° Inclination
11.0	11.0	11.0
17.5	17.5	17.5
21.0	21.0	21.0
57.9	62.2	64.2
142.0	157.0	197.7
198.2	209.5	281.8
217.0	223.7	312.6
1112.2[8]	1180.1[8]	1499.0[8]

Notes:
1) Declination = + 23.5 degrees.
2) Interpolated from Figure 1.
3) Interpolated from Figure 2.
4) Based on an effective transmissivity of 0.605 through two cover glasses.
5) The reflected diffuse radiation was neglected, because it is estimated to be less than 5 percent of the diffuse radiation incident on the collector surface, i.e., $< 0.05\ G_d$.
6) Based on a mirror surface reflectivity of 0.95.
7) The reduction in shortwave radiation to the collector that results from mirror shadowing was neglected.
8) Total flux per day (Btu/ft²·day).

Total Daily Shortwave Radiation (Direct & Diffuse) Transmitted Through the ISC Solar Furnace Covers at 40° North Latitude on a Cloudless Spring or Fall Equinox Day[1]

Local Mean Sun Time		Sun Position (Deg)			Collector only										
					Direct Solar Radiation						Diffuse Sky Radiation				Total
AM	PM	Azi	Alt	Zen	G_n	cos i	G_i	i	τ_i[2]	G_I	G_h	ϵ[3]	G_d	G_D[4]	$G_I + G_D$
6	6	—	0.0	90.0	0.0	0.0	0.0	—	0.0	0.0	0.0	0.0	0.0	0.0	0.0
7	5	80.2	11.4	78.6	152.2	0.243	37.0	80.7	0.23	8.5	30.1	1.50	13.4	8.1	16.6
8	4	69.6	22.5	67.5	254.8	0.470	119.8	69.8	0.54	64.7	97.5	2.50	26.0	15.7	80.4
9	3	57.3	32.8	57.2	299.6	0.664	199.0	57.3	0.69	137.3	162.3	3.30	32.8	19.8	157.1
10	2	42.0	41.6	48.4	321.8	0.813	261.7	43.3	0.75	196.3	213.7	3.96	36.0	21.8	218.1
11	1	22.6	47.7	42.3	332.4	0.908	301.8	28.4	0.76	229.4	245.9	4.38	37.4	22.6	252.0
Noon		0.0	50.0	40.0	335.7	0.940	315.5	20.0	0.77	242.9	257.2	4.55	37.7	22.8	265.7
									Total (Btu/ft²•day) =	1515.3				198.8	1714.1

Mirror Contribution (Direct Solar Radiation only)[5]																		
	0° Inclination						10° Inclination						20° Inclination					
F_a	G_m	F_x	G_r[6]	k	τ_k	G_R	G_m	F_x	G_r[6]	k	τ_k	G_R	G_m	F_x	G_r[6]	k	τ_k	G_R
0.0	0.0	1.00	0.0	—	0.0	0.0	0.0	1.00	0.0	—	0.0	0.0	0.0	1.00	0.0	—	0.0	0.0
0.26	30.1	1.00	7.4	82.7	0.19	1.4	3.7	1.00	0.9	80.9	0.23	0.2	0.0	1.00	0.0	80.2	0.25	0.0
0.55	97.5	1.00	50.9	77.7	0.33	16.8	55.1	1.00	28.8	72.9	0.47	13.5	11.1	1.00	5.8	70.1	0.53	3.1
0.74	162.3	0.84	95.8	75.7	0.39	37.4	116.1	1.00	81.6	66.6	0.59	48.1	66.4	1.00	46.7	60.1	0.66	30.8
0.85	213.7	0.48	82.8	76.4	0.37	30.6	168.6	1.00	136.1	62.5	0.64	87.1	118.5	1.00	95.7	50.7	0.72	68.9
0.93	245.9	0.29	63.0	78.7	0.30	18.9	203.3	0.87	156.3	60.4	0.66	103.2	154.5	1.00	136.5	43.1	0.74	101.0
1.00	257.2	0.23	56.2	80.0	0.26	14.6	215.8	0.78	159.9	60.0	0.67	107.1	167.9	1.00	159.5	40.0	0.75	119.6
Total (Btu/ft²•day) =						224.8						611.3						527.2
Percentage Improvement (%) =						13.1						35.7						30.8

Total Shortwave Radiation[7] $G_I + G_D + G_R$		
0° Inclination	10° Inclination	20° Inclination
0.0	0.0	0.0
18.0	16.8	16.6
97.2	93.9	83.5
194.5	205.2	187.9
248.7	305.2	287.0
270.9	355.2	353.0
280.3	372.8	385.3
1938.9[8]	2325.4[8]	2241.3[8]

Notes:
1) Declination = 0 degrees.
2) Interpolated from Figure 1.
3) Interpolated from Figure 2.
4) Based on an effective transmissivity of 0.605 through two cover glasses.
5) The reflected diffuse radiation was neglected, because it is estimated to be less than 5 percent of the diffuse radiation incident on the collector surface, i.e., $< 0.05\ G_d$.
6) Based on a mirror surface reflectivity of 0.95.
7) The reduction in shortwave radiation to the collector that results from mirror shadowing was neglected.
8) Total flux per day (Btu/ft²·day).

Total Daily Shortwave Radiation (Direct & Diffuse) Transmitted Through the ISC Solar Furnace Covers at 40° North Latitude on a Cloudless Winter Solstice Day[1]

Local Mean Sun Time		Sun Position (Deg)			Collector only										Total
					Direct Solar Radiation						Diffuse Sky Radiation				
AM	PM	Azi	Alt	Zen	G_n	cos i	G_i	i	τ_i[2]	G_I	G_h	ϵ[3]	G_d	G_D[4]	$G_I + G_D$
6	6	—	0.0	90.0	0.0	0.0	0.0	—	0.0	0.0	0.0	0.0	0.0	0.0	0.0
7	5	62.4	0.0	90.0	0.0	0.0	0.0	66.3	0.60	0.0	0.0	0.0	0.0	0.0	0.0
8	4	52.9	5.4	84.6	47.1	0.567	26.7	56.7	0.71	19.0	4.4	0.88	3.4	2.1	21.1
9	3	41.9	13.9	76.1	183.9	0.746	137.2	44.3	0.74	101.5	44.2	1.76	16.7	10.1	111.6
10	2	29.3	20.6	69.4	242.8	0.883	214.4	30.6	0.76	162.9	85.4	2.35	24.3	14.7	177.6
11	1	15.2	25.0	65.0	268.5	0.969	260.1	16.0	0.77	200.3	113.5	2.70	28.0	16.9	217.2
Noon		0.0	26.5	63.5	275.6	0.998	275.1	3.5	0.77	211.8	123.0	2.82	29.1	17.6	229.4
					Total (Btu/ft²•day) = 1179.2									105.2	1284.4

Mirror Contribution (Direct Solar Radiation only)[5]																		
	0° Inclination						10° Inclination						20° Inclination					
F_a	G_m	F_x	G_r[6]	k	τ_k	G_R	G_m	F_x	G_r[6]	k	τ_k	G_R	G_m	F_x	G_r[6]	k	τ_k	G_R
0.0	0.0	1.00	0.0	—	0.0	0.0	0.0	1.00	0.0	—	0.0	0.0	0.0	1.00	0.0	—	0.0	0.0
0.68	0.0	1.00	0.0	—	0.0	0.0	0.0	1.00	0.0	—	0.0	0.0	0.0	1.00	0.0	—	0.0	0.0
0.78	4.4	1.00	3.3	60.5	0.66	2.2	0.0	1.00	0.0	54.4	0.71	0.0	0.0	1.00	0.0	53.0	0.71	0.0
0.85	44.2	1.00	35.7	57.6	0.69	24.6	12.5	1.00	10.1	47.1	0.74	7.5	0.0	1.00	0.0	42.0	0.75	0.0
0.91	85.4	1.00	73.8	56.4	0.69	50.9	44.7	1.00	38.6	41.4	0.75	29.0	2.5	1.00	2.2	31.0	0.76	1.7
0.95	113.5	1.00	102.4	56.4	0.69	70.7	69.5	1.00	62.7	37.8	0.75	47.0	23.4	1.00	21.1	21.2	0.77	16.2
1.00	123.0	1.00	116.9	56.5	0.69	80.6	78.3	1.00	74.4	36.5	0.76	56.5	31.2	1.00	29.6	16.5	0.77	22.8
Total (Btu/ft²•day) =						377.4						223.5						58.6
Percentage Improvement (%) =						29.4						17.4						4.6

Total Shortwave Radiation[7] $G_I + G_D + G_R$		
0° Inclination	10° Inclination	20° Inclination
0.0	0.0	0.0
0.0	0.0	0.0
23.3	21.1	21.1
136.2	119.1	111.6
228.5	206.6	179.3
287.9	264.2	233.4
310.0	285.9	252.2
1661.8[8]	1507.9[8]	1343.0[8]

Notes:
1) Declination = − 23.5 degrees.
2) Interpolated from Figure 1.
3) Interpolated from Figure 2.
4) Based on an effective transmissivity of 0.605 through two cover glasses.
5) The reflected diffuse radiation was neglected, because it is estimated to be less than 5 percent of the diffuse radiation incident on the collector surface, i.e., $< 0.05\ G_d$.
6) Based on a mirror surface reflectivity of 0.95.
7) The reduction in shortwave radiation to the collector that results from mirror shadowing was neglected.
8) Total flux per day (Btu/ft²·day).

Total Daily Shortwave Radiation (Direct & Diffuse) Transmitted Through the ISC Solar Furnace Covers at 45° North Latitude on a Cloudless Summer Solstice Day[1]

Local Mean Sun Time		Sun Position (Deg)			Collector only										
					Direct Solar Radiation						Diffuse Sky Radiation				Total
AM	PM	Azi	Alt	Zen	G_n	cos i	G_i	i	τ_i[2]	G_I	G_h	ϵ[3]	G_d	G_D[4]	$G_I + G_D$
6	6	107.1	16.4	73.6	209.6	0.0	0.0	—	0.0	0.0	59.2	1.99	19.8	12.0	12.0
7	5	97.5	26.7	63.3	275.5	0.112	30.9	—	0.0	0.0	124.3	2.82	29.4	17.8	17.8
8	4	86.8	37.3	52.7	312.2	0.341	106.6	86.8	0.08	8.5	189.2	3.64	34.7	21.0	29.5
9	3	74.9	47.8	42.2	332.6	0.522	173.6	75.6	0.36	62.5	246.4	4.41	37.3	22.6	85.1
10	2	58.6	57.5	32.5	344.3	0.664	228.7	62.5	0.64	146.4	290.4	5.05	38.4	23.2	169.6
11	1	34.6	65.3	24.7	350.5	0.752	263.6	47.8	0.74	195.1	318.4	5.45	39.0	23.6	218.7
Noon		0.0	68.5	21.5	352.4	0.783	275.8	38.5	0.76	209.6	327.9	5.57	39.3	23.8	233.4
							Total (Btu/ft²•day) =			1034.6				264.2	1298.8

Mirror Contribution (Direct Solar Radiation only)[5]																		
	0° Inclination						10° Inclination						20° Inclination					
F_a	G_m	F_x	G_r[6]	k	τ_k	G_R	G_m	F_x	G_r[6]	k	τ_k	G_R	G_m	F_x	G_r[6]	k	τ_k	G_R
0.0	59.2	1.00	0.0	—	0.0	0.0	23.4	1.00	0.0	—	0.0	0.0	0.0	1.00	0.0	—	0.0	0.0
0.0	124.3	1.00	0.0	—	0.0	0.0	79.5	1.00	0.0	—	0.0	0.0	32.3	1.00	0.0	—	0.0	0.0
0.0	189.2	0.64	0.0	88.8	0.03	0.0	143.2	1.00	0.0	87.8	0.07	0.0	92.8	1.00	0.0	87.2	0.07	0.0
0.40	246.4	0.29	27.2	86.8	0.09	2.4	203.9	0.87	67.4	82.0	0.21	14.2	155.1	1.00	58.9	78.1	0.32	18.8
0.73	290.4	0.05	10.1	88.7	0.03	0.3	253.8	0.52	91.5	78.5	0.31	28.4	209.6	1.00	145.4	69.4	0.55	80.0
0.89	318.4	0.0	0.0	—	0.0	0.0	288.2	0.31	75.5	77.9	0.32	24.2	249.1	0.80	168.5	62.1	0.65	109.5
1.00	327.9	0.0	0.0	—	0.0	0.0	300.5	0.23	65.7	78.5	0.31	20.4	263.9	0.70	175.5	58.5	0.68	119.3
	Total (Btu/ft²•day) =					5.4						154.0						535.9
	Percentage Improvement (%) =					0.4						11.9						41.3

Total Shortwave Radiation[7] $G_I + G_D + G_R$		
0° Inclination	10° Inclination	20° Inclination
12.0	12.0	12.0
17.8	17.8	17.8
29.5	29.5	29.5
87.5	99.3	103.9
169.9	198.0	249.6
218.7	242.9	328.2
233.4	253.8	352.7
1304.2[8]	1452.8[8]	1834.7[8]

Notes:
1) Declination = + 23.5 degrees.
2) Interpolated from Figure 1.
3) Interpolated from Figure 2.
4) Based on an effective transmissivity of 0.605 through two cover glasses.
5) The reflected diffuse radiation was neglected, because it is estimated to be less than 5 percent of the diffuse radiation incident on the collector surface, i.e., $< 0.05\ G_d$.
6) Based on a mirror surface reflectivity of 0.95.
7) The reduction in shortwave radiation to the collector that results from mirror shadowing was neglected.
8) Total flux per day (Btu/ft²·day).

Total Daily Shortwave Radiation (Direct & Diffuse) Transmitted Through the ISC Solar Furnace Covers at 45° North Latitude on a Cloudless Spring or Fall Equinox Day[1]

Local Mean Sun Time		Sun Position (Deg)			Collector only										Total
					Direct Solar Radiation						Diffuse Sky Radiation				
AM	PM	Azi	Alt	Zen	G_n	cos i	G_i	i	τ_i[2]	G_I	G_h	ϵ[3]	G_d	G_D[4]	$G_I + G_D$
6	6	—	0.0	90.0	0.0	0.0	0.0	—	0.0	0.0	0.0	0.0	0.0	0.0	0.0
7	5	79.2	10.5	79.5	139.1	0.253	35.3	79.8	0.25	8.8	25.3	1.45	11.7	7.1	15.9
8	4	67.8	20.7	69.3	243.5	0.504	122.7	68.1	0.55	67.5	86.1	2.35	24.4	14.8	82.3
9	3	54.7	30.0	60.0	290.0	0.683	198.2	54.7	0.71	140.7	145.0	3.08	31.4	19.0	159.7
10	2	39.3	37.8	52.2	313.4	0.836	262.0	39.9	0.75	196.5	192.1	3.68	34.8	21.1	217.6
11	1	20.8	43.1	46.9	324.7	0.933	302.9	24.4	0.77	233.2	221.9	4.06	36.4	22.0	255.2
Noon		0.0	45.0	45.0	328.1	0.966	316.9	15.0	0.77	244.0	232.0	4.20	36.8	22.3	266.3
					Total (Btu/ft²•day) = 1537.4									190.3	1727.7

	Mirror Contribution (Direct Solar Radiation only)[5]																	
	0° Inclination						10° Inclination						20° Inclination					
F_a	G_m	F_x	G_r[6]	k	τ_k	G_R	G_m	F_x	G_r[6]	k	τ_k	G_R	G_m	F_x	G_r[6]	k	τ_k	G_R
0.0	0.0	1.00	0.0	—	0.0	0.0	0.0	1.00	0.0	—	0.0	0.0	0.0	1.00	0.0	—	0.0	0.0
0.29	25.3	1.00	7.0	81.8	0.21	1.5	1.2	1.00	0.3	79.9	0.26	0.1	0.0	1.00	0.0	79.2	0.29	0.0
0.59	86.1	1.00	48.3	76.2	0.37	17.9	45.2	1.00	25.3	71.0	0.51	12.9	3.0	1.00	1.7	68.2	0.57	1.0
0.76	145.0	1.00	104.7	73.2	0.45	47.1	99.2	1.00	71.6	63.7	0.63	45.1	50.4	1.00	36.4	57.1	0.69	25.1
0.86	192.1	0.62	97.3	73.0	0.46	44.8	146.2	1.00	119.4	58.7	0.68	81.2	95.8	1.00	78.3	46.8	0.74	57.9
0.94	221.9	0.43	85.2	74.2	0.43	36.6	177.3	1.00	158.3	55.9	0.70	110.8	127.4	1.00	113.8	38.5	0.75	85.4
1.00	232.0	0.37	81.5	75.0	0.41	33.4	188.2	1.00	178.8	55.0	0.70	125.2	138.7	1.00	131.8	35.0	0.76	100.2
Total (Btu/ft²•day) =						329.2						625.4						439.0
Percentage Improvement (%) =						19.1						36.2						25.4

Total Shortwave Radiation[7] $G_I + G_D + G_R$		
0° Inclination	10° Inclination	20° Inclination
0.0	0.0	0.0
17.4	16.0	15.9
100.2	95.2	83.3
206.8	204.8	184.8
262.4	298.8	275.5
291.8	366.0	340.6
299.7	391.5	366.5
2056.9[8]	2353.1[8]	2166.7[8]

Notes:
1) Declination = 0 degrees.
2) Interpolated from Figure 1.
3) Interpolated from Figure 2.
4) Based on an effective transmissivity of 0.605 through two cover glasses.
5) The reflected diffuse radiation was neglected, because it is estimated to be less than 5 percent of the diffuse radiation incident on the collector surface, i.e., $< 0.05\ G_d$.
6) Based on a mirror surface reflectivity of 0.95.
7) The reduction in shortwave radiation to the collector that results from mirror shadowing was neglected.
8) Total flux per day (Btu/ft²·day).

Total Daily Shortwave Radiation (Direct & Diffuse) Transmitted Through the ISC Solar Furnace Covers at 45° North Latitude on a Cloudless Winter Solstice Day[1]

Local Mean Sun Time		Sun Position (Deg)			Collector only										Total
					Direct Solar Radiation						Diffuse Sky Radiation				
AM	PM	Azl	Alt	Zen	G_n	cos i	G_i	i	τ_i[2]	G_I	G_h	ϵ[3]	G_d	G_D[4]	$G_I + G_D$
6	6	—	0.0	90.0	0.0	0.0	0.0	—	0.0	0.0	0.0	0.0	0.0	0.0	0.0
7	5	62.4	0.0	90.0	0.0	0.0	0.0	66.3	0.60	0.0	0.0	0.0	0.0	0.0	0.0
8	4	52.6	2.4	87.6	2.9	0.546	1.6	57.4	0.71	1.1	0.1	0.45	0.2	0.1	1.2
9	3	41.2	10.2	79.8	134.5	0.730	98.2	44.9	0.74	72.7	23.8	1.40	11.3	6.8	79.5
10	2	28.5	16.2	73.8	207.7	0.870	180.8	31.4	0.76	137.4	57.9	1.96	19.7	11.9	149.3
11	1	14.6	20.1	69.9	239.4	0.959	229.5	17.6	0.77	176.7	82.3	2.32	23.7	14.3	191.0
Noon		0.0	21.5	68.5	248.7	0.989	246.0	8.5	0.77	189.4	91.2	2.43	25.0	15.1	204.5
					Total (Btu/ft²•day) = 965.2									81.3	1046.5

	Mirror Contribution (Direct Solar Radiation only)[5]																	
	0° Inclination						10° Inclination						20° Inclination					
F_a	G_m	F_x	G_r[6]	k	τ_k	G_R	G_m	F_x	G_r[6]	k	τ_k	G_R	G_m	F_x	G_r[6]	k	τ_k	G_R
0.0	0.0	1.00	0.0	—	0.0	0.0	0.0	1.00	0.0	—	0.0	0.0	0.0	1.00	0.0	—	0.0	0.0
0.68	0.0	1.00	0.0	—	0.0	0.0	0.0	1.00	0.0	—	0.0	0.0	0.0	1.00	0.0	—	0.0	0.0
0.78	0.1	1.00	0.1	59.1	0.67	0.1	0.0	1.00	0.0	53.6	0.71	0.0	0.0	1.00	0.0	53.0	0.71	0.0
0.85	23.8	1.00	19.2	54.9	0.71	13.6	0.5	1.00	0.4	45.1	0.74	0.3	0.0	1.00	0.0	41.2	0.75	0.0
0.91	57.9	1.00	50.0	52.5	0.72	36.0	22.4	1.00	19.4	38.0	0.75	14.6	0.0	1.00	0.0	29.1	0.76	0.0
0.96	82.3	1.00	75.1	51.6	0.72	54.0	42.0	1.00	38.3	33.2	0.76	29.1	0.4	1.00	0.4	17.7	0.77	0.3
1.00	91.2	1.00	86.6	51.5	0.72	62.4	49.6	1.00	47.1	31.5	0.76	35.8	6.5	1.00	6.2	11.5	0.77	4.8
Total (Btu/ft²•day) = 269.3												123.8						5.4
Percentage Improvement (%) = 25.3												11.8						0.5

Total Shortwave Radiation[7] $G_I + G_D + G_R$		
0° Inclination	10° Inclination	20° Inclination
0.0	0.0	0.0
0.0	0.0	0.0
1.3	1.2	1.2
93.1	79.8	79.5
185.3	163.9	149.3
245.0	220.1	191.3
266.9	240.3	209.3
1316.3[8]	1170.3[8]	1051.9[8]

Notes:
1) Declination = − 23.5 degrees.
2) Interpolated from Figure 1.
3) Interpolated from Figure 2.
4) Based on an effective transmissivity of 0.605 through two cover glasses.
5) The reflected diffuse radiation was neglected, because it is estimated to be less than 5 percent of the diffuse radiation incident on the collector surface, i.e., $< 0.05\ G_d$.
6) Based on a mirror surface reflectivity of 0.95.
7) The reduction in shortwave radiation to the collector that results from mirror shadowing was neglected.
8) Total flux per day (Btu/ft²·day).

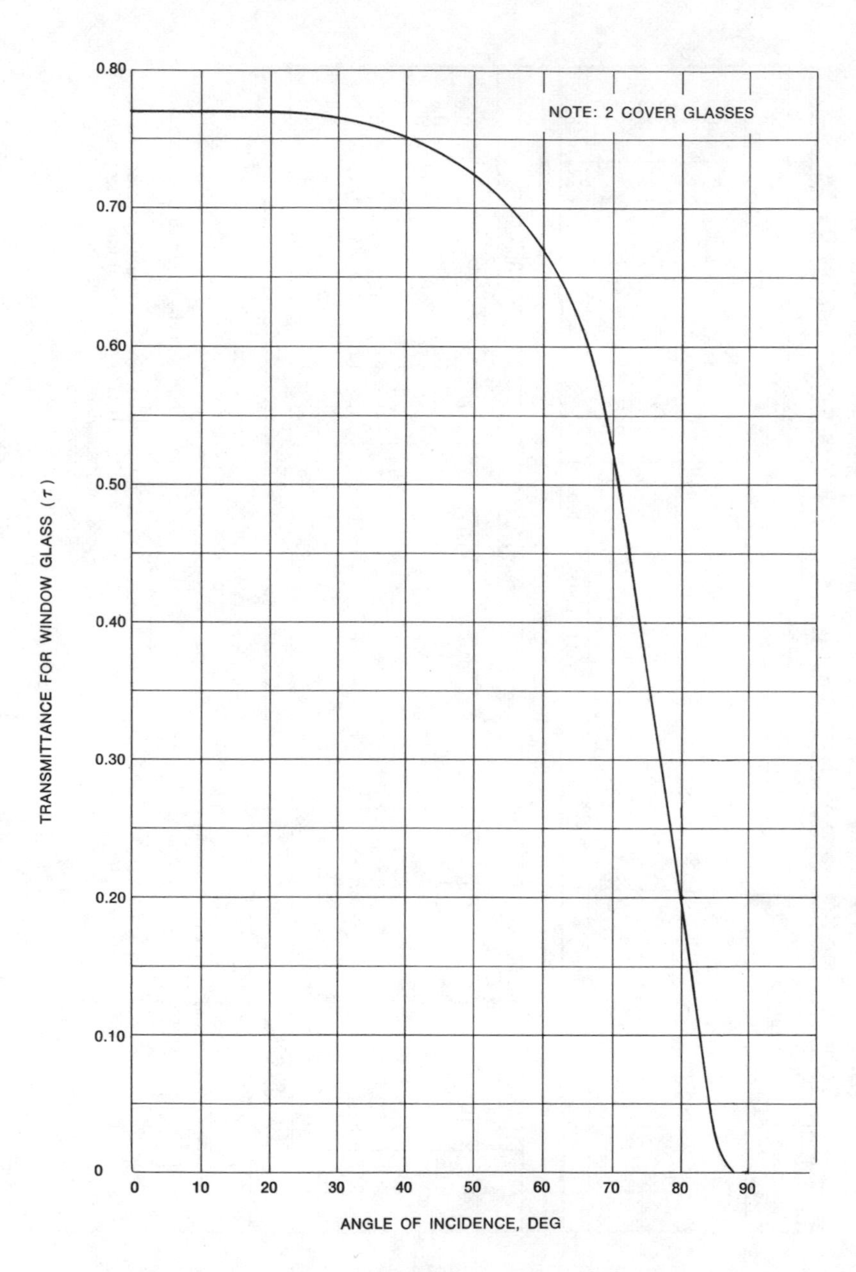

Figure 1. Transmittance of Direct Solar Radiation Through the Glass Covers on the ISC Solar Furnace

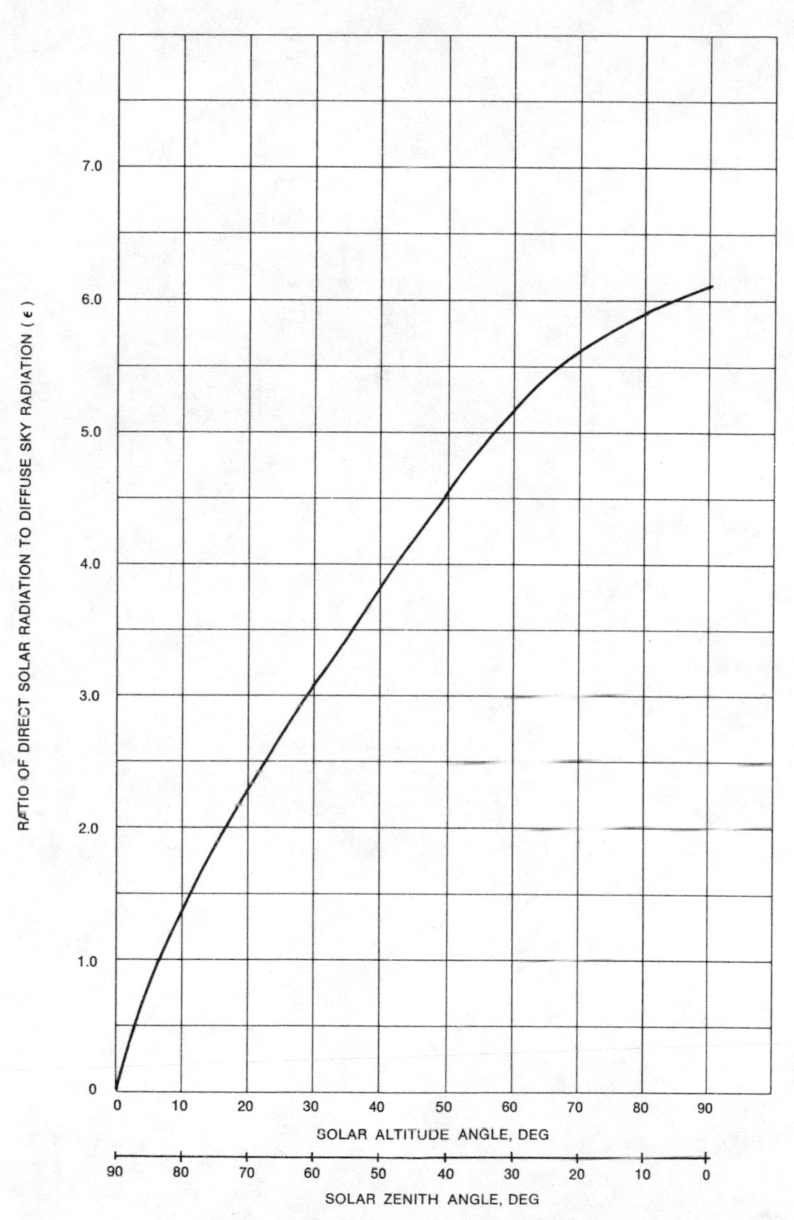

Figure 2. Approximate Ratio of Direct Solar Radiation to Diffuse Sky Radiation on a Horizontal Surface on Cloudless Days

APPENDIX F

METHOD FOR CALCULATING PERFORMANCE CAPABILITY OF THE ISC SOLAR FURNACE USING U.S. WEATHER BUREAU DATA

CONTENTS

CALCULATION PROCEDURE

Step 1. Compute the sun position (solar altitude and azimuth) for each hour of the day, on the 15th day of each month of the heating season.

Step 2. Compute the clear day direct normal radiation (G_n) for each elevation determined in Step 1.

Step 3. Compute the average daily direct solar radiation into the collector under cloudless conditions (G_I) for each month of the heating season.

Step 4. Compute the average daily direct solar radiation reflected by the mirror into the collector under cloudless conditions (G_R) for each month of the heating season.

Step 5. Compute the average daily diffuse sky radiation (G_D) for the locale under consideration, by looking up the measured total shortwave radiation and substracting the direct solar radiation computed in Step 3. Then calculate the diffuse radiation which penetrates the collector covers.

Step 6. Compute the total average daily radiation into the collector by summing the direct, reflected and diffuse components. Then compute the average daily energy into storage by multiplying by collector efficiency. Additionally, in certain installations a reduction will be required to account for thermal losses that may occur during transfer between storage and the house.

Notes: 1) Exact solar altitude and azimuth values may be obtained from *Hydrographic Office Bulletin No. 214. Tables of Computed Altitude and Azimuth,* available from:
Defense Mapping Agency
Hydrology Center
Clearfield, Utah 84016

2) Some references use computations based on the 21st of each month. The 15th of each month is used here, because it is a better approximation of average conditions for the month.

3) Steps 1 through 4 depend only on the latitude of the location for which the calculation is being performed. Thus, the calculations need be completed only once for each latitude.

4) The differences in the relative amounts of direct and diffuse radiation that result from variations in local microclimates do not substantially affect the performance values, since the overall effective transmissivities of direct and diffuse radiation through the collector covers are similar.

SAMPLE CALCULATION FOR COLUMBUS, OHIO (40° NORTH LATITUDE)

The calculation is performed for Columbus, Ohio for which mean daily solar radiation data are available. (See page A69 of this book.)

Step 1. Solar Altitude and Azimuth

Solar altitude (X) and azimuth (A) can be computed using

$$X = \sin^{-1}(\cos D \cos H \cos L + \sin D \sin L)$$

$$A = \sin^{-1}(\cos D \sin H / \cos X)$$

where: D = declination, deg (obtained from an ephemeris)

H = hour angle, deg

L = latitude, deg

Sun Positions for 40° North Latitude

Local Mean Sun Time		Hour Angle (Deg)	Sept. 15 Dec = 3.3°		Oct. 15 Dec = −8.3°		Nov. 15 Dec = −18.3°		Dec. 15 Dec = −23.2°		Jan. 15 Dec = −21.2°	
AM	PM		Alt	Azi	Alt	Azi	Alt	Azi	Alt	Azi	Alt	Azi
6	6	90	2.1	92.5	—	—	—	—	—	—	—	—
7	5	75	13.6	82.8	5.9	73.9	—	—	—	—	—	—
8	4	60	24.8	72.2	16.6	63.4	9.3	56.4	5.7	53.1	7.2	54.5
9	3	45	35.3	59.9	26.3	51.3	18.2	45.0	14.2	42.1	15.8	43.3
10	2	30	44.4	44.3	34.3	36.3	25.3	31.7	20.9	29.5	22.7	30.4
11	1	15	50.9	24.2	39.7	19.5	30.0	16.5	25.3	15.3	27.2	15.7
Noon		0	53.3	0.0	41.7	0.0	31.7	0.0	26.0	0.0	28.8	0.0

Local Mean Sun Time		Hour Angle (Deg)	Feb. 15 Dec = −12.9°		Mar. 15 Dec = −2.4°		Apr. 15 Dec = 9.6°		May 15 Dec = 18.7°	
AM	PM		Alt	Azi	Alt	Azi	Alt	Azi	Alt	Azi
6	6	90	—	—	—	—	6.2	97.4	11.9	104.5
7	5	75	2.9	70.5	9.9	78.4	17.6	87.8	23.2	95.5
8	4	60	13.3	60.2	20.8	67.8	29.0	77.5	34.7	85.9
9	3	45	22.6	48.3	30.9	55.5	39.9	65.3	46.0	74.6
10	2	30	30.2	34.3	39.5	40.3	49.6	49.5	56.6	59.3
11	1	15	35.3	18.0	45.4	21.6	56.8	27.8	65.1	35.6
Noon		0	37.1	0.0	47.6	0.0	59.6	0.0	68.7	0.0

(1) ASHRAE Handbook & Product Directory, 1974 Applications. p. 59.5. eq. 3 & 4.

Step 2. Direct Normal Solar Radiation (G_n) Under Cloudless Conditions

This can be determined from

$$G_n = \frac{A_o C}{e^{(B/\sin X)}} \quad (2)$$

where: A_o = apparent extraterrestrial irradiation at air mass 0, Btu/ft²•hr

B = atmospheric extinction coefficient, dimensionless

C = clearness number, dimensionless

Values of A_o and B are determined from the following table. The exact values of C may be determined from the ASHRAE Handbook. However, the approximate value of C = 1.0 is sufficiently accurate for this calculation.

Month	A_o	B
Sept.	362	0.181
Oct.	376	0.163
Nov.	386	0.151
Dec.	391	0.143
Jan.	392	0.142
Feb.	386	0.143
Mar.	378	0.153
Apr.	364	0.175
May	351	0.195

Direct Normal Solar Radiation at 40° North Latitude

Local Mean Sun Time		Direct Solar Radiation (G_n), Btu/ft²•hr								
AM	PM	Sept.	Oct.	Nov.	Dec.	Jan.	Feb.	Mar.	Apr.	May
6	6	2.6	—	—	—	—	—	—	72.0	136.3
7	5	167.7	77.0	—	—	—	22.9	155.2	204.1	214.0
8	4	235.1	212.5	151.6	92.7	126.3	207.3	245.7	253.7	249.2
9	3	264.7	260.3	238.0	218.3	232.7	266.1	280.6	277.1	267.7
10	2	279.5	281.6	271.1	261.9	271.3	290.5	297.2	289.3	277.9
11	1	286.7	291.3	285.4	279.8	287.3	301.4	304.9	295.3	283.1
Noon		288.8	294.3	289.6	284.7	291.9	304.5	307.3	297.2	284.7

(2) Ibid, p. 59.8, eq. 8.

Step 3. Average Daily Direct Solar Radiation (G_l) Into Collector Under Cloudless Conditions

This may be calculated from

$$G_l = G_n \cos i \, \tau_i \qquad \text{Btu/ft}^2\text{•hr}$$

where: cos i gives the projected area of the collector normal to the sun's rays and is found by

$$\cos i = .5 \sin X + .866 \cos X \cos A$$

τ_i = effective transmittance of the collector covers and is found from the angle of incidence (i) using Figure 1 of Appendix E, p. E18.

The following table summarizes the computation of G_l for Columbus, Ohio.

Direct Solar Radiation Transmitted Through the ISC Solar Furnace Covers at 40° North Latitude

Local Mean Sun Time		Direct Solar Radiation (G_l), Btu/ft²•hr														
		September					October					November				
AM	PM	G_n	cos i	i	τ_i	G_l	G_n	cos i	i	τ_i	G_l	G_n	cos i	i	τ_i	G_l
6	6	2.6	—	—	—	0.0	0.0	—	—	—	0.0	0.0	—	—	—	0.0
7	5	167.7	0.223	77.1	0.30	11.2	77.0	0.290	73.1	0.44	9.8	0.0	—	—	—	0.0
8	4	235.1	0.450	63.3	0.63	66.7	212.5	0.514	59.0	0.68	74.3	151.6	0.554	56.4	0.70	58.8
9	3	264.7	0.643	50.0	0.73	124.2	260.3	0.707	45.0	0.74	136.2	238.0	0.738	42.4	0.75	131.7
10	2	279.5	0.793	37.6	0.75	166.2	281.6	0.858	30.9	0.76	183.6	271.1	0.880	28.4	0.77	183.7
11	1	286.7	0.886	27.6	0.77	195.6	291.3	0.947	18.7	0.77	212.4	285.4	0.969	14.3	0.77	212.9
Noon		288.8	0.918	23.3	0.77	204.1	294.3	0.979	11.7	0.77	221.9	289.6	1.000	1.7	0.77	223.0
		Total (Btu/ft²•day) =				1331.9					1454.5					1397.2

Local Mean Sun Time		Direct Solar Radiation (G_l), Btu/ft²•hr														
		December					January					February				
AM	PM	G_n	cos i	i	τ_i	G_l	G_n	cos i	i	τ_i	G_l	G_n	cos i	i	τ_i	G_l
6	6	0.0	—	—	—	0.0	0.0	—	—	—	0.0	0.0	—	—	—	0.0
7	5	0.0	—	—	—	0.0	0.0	—	—	—	0.0	22.9	0.314	71.7	0.48	3.5
8	4	92.7	0.567	55.5	0.70	36.8	126.3	0.562	55.8	0.70	49.7	207.3	0.534	57.7	0.69	76.4
9	3	218.3	0.746	41.8	0.75	122.1	232.7	0.743	42.0	0.75	129.7	266.1	0.724	43.6	0.74	142.6
10	2	261.9	0.883	28.1	0.77	178.1	271.3	0.882	28.1	0.77	184.3	290.5	0.870	29.0	0.76	192.1
11	1	279.8	0.969	14.3	0.77	208.8	287.3	0.970	14.1	0.77	214.6	301.4	0.961	16.0	0.77	223.0
Noon		284.7	0.998	3.2	0.77	218.8	291.9	1.000	1.3	0.77	224.8	304.5	0.992	7.1	0.77	232.6
		Total (Btu/ft²•day) =				1310.4					1381.4					1507.8

Local Mean Sun Time		Direct Solar Radiation (G_l), Btu/ft²•hr														
		March					April					May				
AM	PM	G_n	cos i	i	τ_i	G_l	G_n	cos i	i	τ_i	G_l	G_n	cos i	i	τ_i	G_l
6	6	0.0	—	—	—	0.0	72.0	—	—	—	0.0	136.3	—	—	—	0.0
7	5	155.2	0.258	75.1	0.37	14.8	204.1	0.183	79.5	0.22	8.2	214.0	0.121	83.1	0.10	2.6
8	4	245.7	0.483	61.1	0.66	78.3	253.7	0.406	66.0	0.60	61.8	249.2	0.336	70.4	0.52	43.5
9	3	280.6	0.678	47.3	0.74	140.8	277.1	0.598	53.2	0.71	117.7	267.7	0.519	58.7	0.68	94.5
10	2	297.2	0.828	34.1	0.76	187.0	289.3	0.745	41.8	0.75	161.6	277.9	0.661	48.6	0.73	134.1
11	1	304.9	0.921	22.9	0.77	216.2	295.3	0.838	33.1	0.76	188.1	283.1	0.750	41.4	0.75	159.2
Noon		307.3	0.953	17.6	0.77	225.5	297.2	0.869	29.6	0.76	196.3	284.7	0.780	38.7	0.75	166.5
		Total (Btu/ft²•day) =				1499.7					1271.1					1034.3

Step 4. Average Daily Direct Solar Radiation Reflected Into the Collector By the Mirror (G_R) Under Cloudiness Conditions

This is given by the formula

$$G_R = \rho \tau_k F_a F_x \cos j \, G_n$$

where: ρ = mirror reflectivity (assumed to be 0.95 for this calculation)

τ_k = effective transmittance of collector covers to reflected light and is found using Figure 1 of Appendix E, p. E18 and the angle of incidence of the reflected light (k) which in turn is determined from

$$k = \cos^{-1}[\cos(120° - 2\phi)\cos Z + \sin(120° - 2\phi)\sin Z \cos A]$$

F_a accounts for reflected light which misses the collector to the edge. It is a function of the length of the solar furnace. For the smallest ISC solar furnace it is

$$F_a = 1 - 1/3\left\{1 + [\cos(60° + \phi)]/f_x) \frac{\tan A}{\cos \phi}\right\} \qquad \text{for } F_a > 0.5$$

$$F_a = .75 f_x \cos\phi/\tan A\,[\cos(60° + \phi) + f_x] \qquad \text{for } F_a < 0.5$$

where: $f_x = \sin(60° + 2\phi - X_s)/\sin(X_s - \phi)$

F_x accounts for the reflected light which misses the collector by passing over the top. It is given by

$$F_x = \sin(60° + 2\phi - X_s)/\sin(X_s - \phi) \qquad \text{for } 30° + 1.5\phi < X_s < 60° + 2\phi$$

$$F_x = 1 \qquad \text{for } X_s < 30° + 1.5\phi$$

where: $X_s = \tan^{-1}(\tan X/\cos A)$

Cos j gives the projected area of the mirror to the sun's rays and is given by

$$\cos j = \cos\phi \cos Z - \sin\phi \sin Z \cos A$$

The optimum mirror tilt is a complicated function of sun position, intensity of sunlight and solar furnace geometry. However, it is approximately the mirror tilt that gives maximum coverage of the collector with reflected light at midday.

$$\phi_{opt} \approx .667\left[\tan^{-1}\left(\frac{\tan X}{\cos A}\right) - 30°\right]$$

Month	ϕ_{opt} (Deg)
Sept.	16
Oct.	8
Nov.	1

Month	ϕ_{opt} (Deg)
Dec.	0
Jan.	0
Feb.	4

Month	ϕ_{opt} (Deg)
Mar.	12
Apr.	20
May	26

Mirror Contribution to the ISC Solar Furnace at 40° North Latitude

Local Mean Sun Time AM	PM	Reflected Direct Solar Radiation (G_R), Btu/ft²•hr — September (16° Mirror Tilt) G_m	F_a	F_x	k	τ_k	G_R	October (8° Mirror Tilt) G_m	F_a	F_x	k	τ_k	G_R	November (1° Mirror Tilt) G_m	F_a	F_x	k	τ_k	G_R
6	6	0.1	—	—	—	—	0.0	0.0	—	—	—	—	0.0	0.0	—	—	—	—	0.0
7	5	32.5	—	0.65	82.7	0.12	0.0	5.0	—	1.00	76.1	0.34	0.0	0.0	—	—	—	—	0.0
8	4	77.3	0.18	0.87	73.4	0.43	5.2	48.0	0.30	1.00	70.0	0.53	7.6	23.5	0.42	1.00	66.3	0.60	5.9
9	3	118.3	0.33	0.94	65.1	0.61	22.5	95.0	0.45	1.00	64.6	0.62	26.5	72.4	0.55	1.00	63.9	0.63	25.3
10	2	149.8	0.58	0.98	58.3	0.68	57.6	132.6	0.69	1.00	60.0	0.67	60.9	113.3	0.71	1.00	61.9	0.65	51.9
11	1	170.0	0.81	1.00	53.8	0.71	97.3	156.4	0.85	1.00	57.4	0.69	91.4	139.8	0.85	1.00	60.7	0.66	78.7
Noon		176.7	1.00	1.00	52.1	0.72	127.2	165.1	1.00	0.98	56.5	0.69	112.1	149.1	1.00	0.96	60.3	0.66	94.7
Total (Btu/ft²•day) =							492.4						484.9						418.3
Mirror Contribution (ρ = .95) =							467.8						460.7						397.4

Local Mean Sun Time AM	PM	Reflected Direct Solar Radiation (G_R), Btu/ft²•hr — December (0° Mirror Tilt) G_m	F_a	F_x	k	τ_k	G_R	January (0° Mirror Tilt) G_m	F_a	F_x	k	τ_k	G_R	February (4° Mirror Tilt) G_m	F_a	F_x	k	τ_k	G_R
6	6	0.0	—	—	—	—	0.0	0.0	—	—	—	—	0.0	0.0	—	—	—	—	0.0
7	5	0.0	—	—	—	—	0.0	0.0	—	—	—	—	0.0	0.6	—	1.00	73.0	0.44	0.0
8	4	9.2	0.51	1.00	62.1	0.65	3.0	15.8	0.47	1.00	64.1	0.62	4.6	39.8	0.35	1.00	68.4	0.56	7.8
9	3	53.5	0.63	1.00	60.0	0.67	22.4	63.3	0.60	1.00	61.9	0.65	24.5	89.4	0.50	1.00	64.3	0.62	27.7
10	2	93.5	0.75	1.00	58.3	0.68	47.5	104.7	0.73	1.00	60.3	0.66	50.3	129.9	0.68	1.00	61.1	0.66	58.4
11	1	119.5	0.87	1.00	57.2	0.69	72.0	131.3	0.87	1.00	59.1	0.67	76.1	155.8	0.84	1.00	59.0	0.68	89.4
Noon		128.4	1.00	1.00	56.8	0.69	88.6	140.7	1.00	1.00	58.8	0.68	95.7	164.4	1.00	0.97	58.3	0.68	108.8
Total (Btu/ft²•day) =							378.4						406.7						475.4
Mirror Contribution (ρ = .95) =							359.5						386.4						451.6

Local Mean Sun Time AM	PM	Reflected Direct Solar Radiation (G_R), Btu/ft²•hr — March (12° Mirror Tilt) G_m	F_a	F_x	k	τ_k	G_R	April (20° Mirror Tilt) G_m	F_a	F_x	k	τ_k	G_R	May (26° Mirror Tilt) G_m	F_a	F_x	k	τ_k	G_R
6	6	0.0	—	—	—	—	0.0	10.4	—	—	—	—	0.0	39.9	—	—	—	—	0.0
7	5	19.9	0.12	1.00	79.8	0.21	0.5	55.5	—	0.33	84.9	0.05	0.0	83.9	—	0.18	86.1	0.02	0.0
8	4	67.8	0.23	1.00	71.9	0.48	7.6	99.2	0.13	0.69	74.3	0.59	5.1	120.9	0.04	0.56	74.3	0.59	1.7
9	3	113.5	0.39	1.00	64.9	0.61	26.9	136.6	0.27	0.87	64.7	0.62	19.8	151.0	0.17	0.79	63.6	0.63	12.8
10	2	149.6	0.62	1.00	59.2	0.67	62.3	165.5	0.51	0.96	56.9	0.69	55.6	173.7	0.37	0.93	54.7	0.71	42.6
11	1	172.2	0.82	1.00	55.5	0.70	99.2	183.4	0.78	1.00	51.5	0.72	103.1	187.7	0.72	1.00	48.5	0.73	98.1
Noon		180.2	1.00	1.00	54.2	0.71	127.7	189.3	1.00	1.00	49.6	0.73	138.2	192.5	1.00	1.00	46.3	0.74	142.5
Total (Btu/ft²•day) =							520.7						505.4						452.9
Mirror Contribution (ρ = .95) =							494.7						480.1						430.3

Step 5. Average Daily Diffuse Sky Radiation (G_D)

The total measured shortwave radiation and the total computed direct radiation must first be stated in the same form. The direct radiation can then be subtracted from the measured value to arrive at the amount of diffuse. The common form chosen is Btu/ft² on a horizontal surface. To convert the measured radiation in langleys to Btu/ft², multiply by 3.687. To convert the direct normal radiation for cloudless conditions to direct radiation on a horizontal surface for average conditions, multiply by the mean percentage of possible sunshine (sun factor) and the sine of the solar altitude (sin X).

To convert the diffuse radiation on a horizontal surface to the diffuse radiation penetrating the collector covers, multiply by the view factor of the collector to the sky (.75) and the transmittance of the collector covers to diffuse radiation (.605).

The following table summarizes the computation of the diffuse radiation input for Columbus, Ohio.

Month	Avg. Daily Total Radiation on a Horizontal Surface	Mean Percentage of Possible Sunshine	Avg. Daily Direct Radiation on a Horizontal Surface	Avg. Daily Diffuse Radiation on a Horizontal Surface	Avg. Daily Diffuse Radiation Penetrating Collector Covers
Sept.	1555.9	0.66	1088.9	467.0	211.9
Oct.	1054.5	0.60	751.9	302.6	137.3
Nov.	648.9	0.44	381.5	267.4	121.3
Dec.	475.6	0.35	238.0	237.6	107.8
Jan.	471.9	0.36	277.6	194.3	88.2
Feb.	737.4	0.44	495.7	241.7	109.7
Mar.	1095.0	0.49	762.1	332.9	151.1
Apr.	1441.6	0.54	1043.1	398.5	180.8
May	1736.6	0.63	1346.0	390.6	177.2

Step 6. Heat Produced by the ISC Solar Furnace in Columbus, Ohio

To determine the heat produced by the ISC solar furnace

1) sum the direct (G_I) and reflected (G_R) inputs
2) multiply the sum by the mean percentage of possible sunshine for the area
3) add the diffuse (G_D) input
4) multiply the total by the collector efficiency

The following table summarizes this computation for Columbus, Ohio.

Month	Direct Solar Radiation Transmitted (Btu/ft²•day)	Sun Factor	Diffuse Solar Radiation Transmitted (Btu/ft²•day)	Collector Efficiency	Heat Produced (Btu/ft²•day)
Sept.	1799.7	0.66	211.9	0.65	909.8
Oct.	1915.2	0.60	137.3	0.82	1054.9
Nov.	1794.6	0.44	121.3	0.90	819.8
Dec.	1669.9	0.35	107.8	0.93	643.8
Jan.	1767.8	0.36	88.2	0.94	681.1
Feb.	1959.4	0.44	109.7	0.94	913.5
Mar.	1994.4	0.49	151.1	0.92	1038.1
Apr.	1751.2	0.54	180.8	0.88	991.3
May	1464.6	0.63	177.2	0.70	769.9

Notice that collector efficiency increases in midwinter, when the heating demand is high and average storage temperature is low. This is a typical operating characteristic of a well-insulated collector. The actual value of collector efficiency is a function of the heat demand of the house on which the solar furnace is installed.

Heat produced means heat transferred to storage. Depending on how well the solar furnace is matched to the house, some of this heat may be lost from storage or from the supply duct between the solar furnace and the house.

APPENDIX G

ADVICE TO THE CONSUMER CONSIDERING THE PURCHASE OF SOLAR HEATING EQUIPMENT

Rising fuel prices and regional shortages of some types of fuel have spurred tremendous interest in the utilization of solar energy for home heating. All over the United States, people are adding solar heating to their existing homes or incorporating it into their new homes.

Unfortunately, as with anything new and in demand, the "blue-suede shoe boys" are flocking to fleece the public.

The following tips are given to help you protect yourself from being bilked. There are many reputable firms with reliable products now in the market place. These common-sense rules may help you proceed.

1. *Hire an independent professional engineer who is registered in your state to advise you!* Reputable firms will have professional engineers on their design staffs, but you should have your own. Just as you should have an attorney advise you when you are purchasing your home, the registered professional engineer can save you literally thousands of dollars. For example, adequate solar insulation in your home can *cut in half* the size collector and storage unit needed – something the salesman may neglect to tell you! Be sure that performance claims have been checked by your professional engineer. (Look under Engineers, Consulting in the Yellow Pages.)

2. *Have your attorney check the sales contract and warranty!* Be sure that your system will have at least a one year warranty on motors, pumps, blowers and controls, and at least five years on structural parts. These warranties should cover *parts and labor*.

3. *Check out the firm with which you are dealing with the Better Business Bureau and Chamber of Commerce*. This is a simple step, but very useful.

4. *Do deposits go into escrow?* Because demand is much greater than supply capabilities in this new field, *it is usual* for dealers to ask for 50% with the order and 50% upon delivery or installation. Are deposits escrowed or bonded until delivery?

5. *Be sure all the costs are covered.* For example, in your new home, if half the basement is taken up by storage, half the cost of that basement (excavation, foundations, concrete, etc.) is part of the cost of that system. Similarly, if the collector is on the roof, a roof which is framed at a 60° angle will cost $2000 to $6000 *more* than a standard roof. This cost is *part of the cost* of the system. Fast-talking salesmen can get you enthusiastic about installing solar heat and neglect to tell you such basics.

6. *Simple test – does the system have blanket certification by your local building department?* If so, the building department will issue a certificate so stating *(but not endorsing the system)*. Beware if you can't see such certification! This means that the system is still in the prototype stage, and you are not necessarily buying a safe and tested system which conforms to the Uniform Building Code.

7. *Insist on seeing a products liability policy from the manufacturer.* Any reputable manufacturer will have a policy which a major insurance company has issued to him against harm to the public from his product – usually in the amount of hundreds of thousands of dollars. If he can't obtain this kind of necessary business insurance on his product, it is a red flag for you!

8. *Performance Warranty.* The reputable manufacturer will have the following performance warranty on his product. "We warrant the product to have the following performance characteristics:

   A. Collection Capability. Cloudless-day heat transfer to storage of at least .... Btu's/day on (specified days of the heating season with solar radiation of .... Btu's/ft^2) and an initial storage temperature of .... °F.

B. Storage Capability. The ratio of useful heat remaining in storage after a 72 hour period of stagnation to useful heat in storage before the period of stagnation shall be at least ....%. (Period of stagnation is defined as a period during which heat is neither collected nor distributed.) Outside ambient temperature should be specified.

If the system is incapable of satisfying this performance during the first ....year(s) after installation, the system shall be replaced or repaired at no charge to the purchaser."

9. *Get an installed price on the contract, unless it is a do-it-yourself kit.* The installed price should include the cost of the building permit, sheet metal, and electrical and plumbing services. If remodeling is involved, repair to drywall, landscaping, etc. should be included. Be sure that glazing is included. The local glazier will charge as much as $1.50 to $6.00 per square foot for the double glass on the collector – more if on a roof-top. Since this can be a major expense, some firms have left this "for you to have done".

10. *Rule of thumb for cost.* Figure that the total system (collector, storage, controls, etc. installed) will cost from $25.00 to $45.00 per square foot of collector area. If the quoted price is less, be suspicious.

11. *Do-it-yourself kits.* Since the installation of any solar heating system is very labor intensive, *major* savings can be effected by the do-it-yourselfer. But before you buy, get a set of assembly instructions. Be sure they are clear and easy to follow. Be sure of just what is included and what is not in the kit. For example, typically a rock storage system will not include the

gravel, since it is not economic to ship. Check the price of the quantity needed delivered at your site from the local sand and gravel company. Many times, glass will not be included – get the delivered price for the glass specified from your local glazier. Expect the warranty on do-it-yourself kits to have these additional words "if installed in a workmanlike manner according to instructions . . ."

12. *Perform solar insulation.* Insulation is much *cheaper* than solar collector and storage systems. Solar insulation will have full 3⅝ in. sidewall and 18 in. ceiling fiberglass batt insulation, with double pane glass and storm doors. Common-sense weatherstripping and caulking are important and well worth the savings. Conservation is not as romantic as solar heating, but accounts for at least 50% of the success of a good solar heating system.

13. *Check the required operating and maintenance costs.* A collector with thin film plastic covers, for example, may be expected to require cover replacement every one or two years. A system using ethylene glycol (antifreeze) will have to have replacement of that fluid at definite periods. What does it cost to run the system?

14. *Check with your insurance agent.* Be sure that you can include your solar heating system in your present homeowner's insurance policy. Find out what, if any, price increase will be involved.

GLOSSARY

Definitions of Technical Terms Used in the Engineering of Solar Heating Systems.

Absorptivity—The capacity of a material to absorb radiant energy. Absorbtance is the ratio of the radiant energy absorbed by a body to that incident on it.

Active Residential Solar Heating System—A solar heating system for heating homes that utilizes forced circulation of the collection and distribution transfer media. It is a system that combines the means for collecting, controlling, transporting and storing solar energy with the primary heating system in the house.

Adiabatic—The process in which there is no heat flow between a substance or system and its surroundings.

Air Mass—The path length of solar radiation through the earth's atmsphere considering the vertical path at sea level as unity.

Albedo—See the definition of reflectance under *Reflectivity*.

Angstrom—A unit of measurement of length equivalent to 1 x 10^{-4} micron.

Apparent Extraterrestial Irradiation at Air Mass Zero—A variable, expressed in Btu/$ft^2\cdot$ hr, used to calculate the direct normal solar radiation incident at the earth's surface on a clear day. It accounts for the seasonal variation of the distance between the earth and sun.

Atmospheric Extinction Coefficient—A dimensionless variable used to calculate the direct normal solar radiation incident at the earth's surface on a clear day. It accounts for the seasonal variation of the water vapor content of the atmosphere.

Baffle—A surface used for deflecting fluids, usually in the form of a plate or wall.

Black Body—A body that absorbs all incident radiation and reflects or transmits none. Additionally, a black body is a perfect radiator, i.e. it emits or radiates the maximum amount of radiant energy for any surface at any given temperature.

British Thermal Unit (Btu)—A unit of measurement equivalent to the amount of heat energy required to raise the temperature of one pound of water one degree Fahrenheit.

Clearness Number—The ratio between the actual clear-day direct solar radiation intensity at a specific location and the intensity calculated for the standard atmosphere for the same location and date.

Coefficient of Linear Expansion—The change in length per unit length per degree change in temperature.

Collection Circuit—The path followed by the collection transfer medium as it removes heat from the collector and transfers it to storage.

Collector Efficiency—The ratio of the amount of heat usefully transferred from the collector into storage to the total solar radiation transmitted through the collector covers. Note: Some authorities define collector efficiency as the ratio of the amount of heat usefully collected to the total solar radiation *incident* on the collector.

Conductance, Thermal—The time rate of heat flow through a body per unit area for a unit temperature difference between the body's surfaces under steady-state conditions.

Conductivity, Thermal—The time rate of heat flow through a homogeneous material per unit area and thickness, under steady-state conditions, when a unit temperature gradient is maintained in the direction normal to the cross-sectional area.

Continuous Air Circulation (CAC)—A mode of operation of a forced-air heating system in which the blower operates continually.

Controller, Automatic—A device used to regulate a system on the basis of its response to changes in the magnitude of some property of the system, e.g. pressure, temperature, etc.

Damper—A device used to vary the volume of air passing through an air outlet, inlet or duct.

Degree Day (DD)—A unit of measurement, based on temperature difference and time, used in estimating average heating requirements for a building. For any one day, when the mean outside temperature is less than 65°F, there exist as many DD as there are Fahrenheit degrees difference in temperature between the mean temperature and 65°F. The base of 65°F assumes that no heat input is required to maintain the inside temperature at 70°F when the outside temperature is 65°F.

Density—The ratio of the mass of a substance to its volume.

Design Outside Temperature—The lowest temperature which usually occurs during the heating season at a given location. This temperature is approximately 15°F above the lowest temperature ever recorded by the meteorological station in the area.

Diffusivity, Thermal—The property of a material equivalent to its thermal conductivity divided by the product of its density and specific heat.

Distribution Circuit—The path followed by the distribution transfer medium as it removes heat from storage and transfers it to the house.

Downpoint Temperature—The temperature of the distribution transfer medium as it leaves storage below which useful heat cannot be delivered.

Drawdown—The removal of all useful heat from storage.

Duct—A passageway used for transporting air or other gas at low pressures.

Electromagnetic Spectrum—The arrangement of electromagnetic radiations (e.g. infrared, visible, ultraviolet, etc.) on a wavelength or frequency scale.

Emissivity—The capacity of a material to emit radiant energy. Emittance is the ratio of the total radiant energy emitted by a body to that emitted by a black body at the same temperature. Note: The emissivity of a surface is numerically equal to its absorptivity when the radiating source is a black body at the same temperature as the surface.

Eutectic—A mixture of two or more substances that melts or freezes at constant temperature and with constant composition. The term is usually applied to that mixture of the given substance which has the lowest melting point.

Equinox—The two times of the year when the sun crosses the equator, thereby making day and night of equal length. The spring equinox occurs about March 21 and the fall equinox about September 21.

Filter—A device used to remove solid materials from a fluid.

Fin—An extended surface used to increase the heat transfer area.

Fluid—A gas, vapor or liquid.

Forced Air Heating System—A heating system in which air, circulated mechanically by either a blower or fan, is the transfer medium.

Heat—The form of energy transferred from one mass to another by virtue of a temperature difference.

Heat Capacity—The quantity of heat required to raise the temperature of a given mass of a substance one degree.

Heat Exchange Flow Pattern—The relative flow arrangement of the collection and distribution circuits in storage. The typical patterns are:

1. Counterflow—collection and distribution transfer media flow in opposite directions through storage; used to enhance stratification.
2. Parallel—collection and distribution transfer media flow in the same direction through storage.

Heat of Fusion (Phase Change) Storage—A heat storage medium in which the addition or removal of heat results in the medium changing state (between solid and liquid) at a constant temperature.

Heat Pump—A refrigerating system installed where the heat discharged from the condenser is desired rather than the heat absorbed by the evaporator.

Heat Transfer—The methods by which heat may be propagated or conveyed from one place to another. They are:

1. Conduction—heat is transferred from one part of a body to another part of the same body or from one body to another in physical contact with it without displacement of the matter within the body.
2. Convection—heat is transferred from one point to another by being carried along as internal energy with the flowing medium which can be either a gas, vapor or liquid. The two types are:
   a. Forced—results from forced circulation of a fluid, as by a fan or pump.
   b. Natural—caused by differences in density resulting from temperature changes.
3. Radiation—heat is transferred from one body to another by the passage of radiant energy between the two. The radiant energy is then converted back to internal energy when it is absorbed by the receiving body.
   a. Longwave—radiant energy emitted from bodies at wavelengths longer than 3.0 microns.
   b. Shortwave—see *Radiation, Solar*

Heat Transfer Coefficient—The unit-surface thermal conductance for convection or radiation which describes the rate by which heat can be transferred by either mode.

Hydronic Heating System—A heating system in which water is the transfer medium.

Infiltration—Air flowing inward as through a wall, crack, etc.

Insolation—The solar radiation incident at the earth's surface.

Insulation—A material having a relatively high resistance to heat flow and used principally to retard the flow of heat.

Irradiation—see *Radiation, Incident*

Langley—A unit of measurement of solar radiation equal to one gram calorie per square centimeter.

Latent Heat—Heat which is necessary to produce a change of state of a material at a constant temperature.

Micron—A unit of measurement of length equivalent to one millionth of a meter.

Nameplate Rating—A statement by the manufacturer of a solar heating system which gives the performance of the system under specific operating conditions.

Net Radiometer—A radiation balance meter used to measure all radiation (both shortwave and longwave) components.

Overall Coefficient of Heat Transfer—The time rate of heat flow through a body per unit area for a unit temperature difference between the fluids on the two sides of the body under steady-state conditions.

Pressure—The force exerted by a substance on a unit area of its boundary.

Primary Heating System—A system used to heat the house when the solar heating system cannot provide useful heat, usually a conventional oil, gas or electric furnace or heat pump.

Pyranometer—An instrument used to measure the total hemispherical solar radiation incident on a surface. This includes direct radiation from the sun, diffuse radiation from the sky and reflected shortwave radiation (albedo) from the surroundings.

Pyrheliometer—An instrument used to measure direct solar radiation incident on a surface located normal to the sun's rays.

Radiant Energy—The energy in the form of electromagnetic waves which is continually emitted from the surface of all bodies.

Radiation, Incident—The quantity of radiant energy incident on a surface per unit time and unit area.

Radiation, Infrared (IR)—Radiant energy of wavelenghts longer than those corresponding to red light, i.e. longer than approximately 0.8 microns.

Radiation, Solar—Radiant energy emitted from the sun in the wavelength range between 0.3 and 3.0 microns. Of the total solar radiation reaching the earth, approximately 3% is in ultraviolet region, 44% in the visible region and 53% in the infrared region.

1. Diffuse Sky—solar radiation received from the sun after its direction has been changed by reflection and scattering by the atmosphere.
2. Direct (Beam)—solar radiation received from the sun without undergoing a change of direction.

Radiation, Thermal—See *Radiant Energy*

Radiation, Ultraviolet (UV)—Radiant energy of wavelengths from 0.1 to 0.4 microns.

Radiation, Visible—Radiant energy of wavelengths from 0.4 to 0.76 microns which produces a sensation defined as seeing when it strikes the retina of the human eye.

Reflectivity—The capacity of a material to reflect radiant energy. Reflectance is the ratio of the radiant energy reflected from a body to that incident on it.

Reflector—A mirror used to increase shortwave radiation input into the collector.

Resistance, Thermal—The reciprocal of thermal conductance (see *Conductance, Thermal).*

Resistance, Temperature Detector (RTD)—A temperature measuring device that employs a sensitive element of extremely pure platinum, copper or nickel wire which provides a definite value at each temperature within its range.

Retro-fit—The installation of a solar heating system on an existing building.

Selective Surface—A Solar collecting surface consisting of a thin coating having high absorptance for solar radiation and a substrate with low emittance for longwave radiation.

Sensible Heat—Heat which produces a change of temperature in a body.

Sensible Heat Storage—A heat storage medium in which the addition or removal of heat results in temperature change only (as opposed to phase change, chemical reaction, etc.). The medium is typically water or gravel (pebble bed).

Solar Altitude—The angle of the sun above the horizon.

Solar Azimuth—The horizontal angle between the sun and due south.

Solar (Photovoltaic) Cell—A device made from semiconductor materials which absorbs solar radiation and converts it into electrical energy.

Solar Collector—A device used to collect solar radiation and convert it to heat. There are two broad categories:

1. Fixed Plate—stationary and does not concentrate the solar radiation, i.e., the absorbing area is the same size as the area intercepting the incoming radiation.
2. Concentrating (Focusing)—concentrates the solar radiation incident on the total area of the reflector onto an absorbing surface of smaller area, thereby increasing the energy flux.

Solar Constant—The amount of solar radiation incident on a unit area of surface located normal to the sun's rays outside of the earth's atmosphere at the earth's mean distance from the sun.

Solar Declination—The angle of the sun north or south of the equatorial plane. It is plus if north of the plane and minus if south.

Solar Furnace—A unitized, self-contained, solar heating system.

Solar Noon—The time of day when the sun is due south, i.e. when the solar azimuth is zero and the solar altitude a maximum.

Solstice—The two times of the year when the sun is furthest north or south of the equator. In the northern hemisphere, the summer solstice occurs about June 21 and the winter solstice about December 21.

Specific Heat—The amount of heat that has to be added to or taken from a unit of weight of a material to produce a change of one degree in its temperature.

Stagnation—A condition in which heat is not added to or removed from storage mechanically, but only as a result of natural heat transfer from the storage container.

Statis—See *Thermal Equilibrium.*

Stefan-Boltzmann Constant—The numerical constant for a perfect radiator (black body). It has a value of 0.173×10^{-8} Btu/hr·ft^2 °R^4.

Stratification—The existence of persistent temperature gradients in storage mediums.

Sun Factor—The average amount of solar radiation measured divided by the total possible solar radiation for a given month and location. This number indicates the amount of cloudiness that occurs at a given location.

Temperature—A measure of heat intensity or the ability of a body to transmit heat to a cooler body.

Thermal Equilibrium—The state of a system at which there are no variations in temperature from one point to another in the system.

Thermal Lag—The amount of heat necessary to reachieve downpoint temperature after collection is resumed following a period of stagnation.

Thermistor—A temperature measuring device that employs a resistor with a high negative temperature coefficient of resistance. As the temperature increases, the resistance goes down and vica versa.

Thermocouple—A terperature measuring device that utilizes the principle that an electromotive force is generated whenever two junctions of two dissimilar metals in an electrical circuit are at different temperatures.

Thermostat—An instrument that controls temperature by responding to changes in temperature.

Transfer Medium—The substance(s) that carries heat from the collector to storage and from storage to the house. The medium is typically a fluid such as air, water or a water-ethylene glycol solution.

Transmissivity—The capacity of a material to transmit radiant energy. Transmittance is the ratio of the radiant energy transmitted through a body to that incident on it.

Useful Heat—Heat delivered by the solar system on demand by the thermostat that contributes to a reduction in the conventional fuel heating bill. Useful solar heat replaces an equivalent amout of heat that otherwise would have to be provided by the primary heating system.

Vapor Barrier—A moisture impervious layer applied to prevent moisture from traveling to a point where it may condense due to lower temperatures.

Viscosity—The internal resistance of fluids to shear.